PENGUIN BOOKS

FINITE AND INFINITE GAMES

'There are at least two kinds of games. One could be called finite, the other infinite. A finite game is played for the purpose of winning, an infinite game for the purpose of continuing the play.'

'The rule of a finite game may not change; the rules of an infinite game must change.'

'Finite players play within boundaries; infinite players play with boundaries.'

'Finite players are serious; infinite players are playful.'

'Finite players win titles; infinite players have nothing but their names.'

'A finite player plays to be powerful; an infinite player plays with strength.'

'Finite players are theatrical; infinite players are dramatic.'

'A finite player consumes time; an infinite player generates time.'

'The finite player aims to win eternal life; the infinite player aims for eternal birth.'

About the author

James P. Carse is Professor of Religion at New York University. A winner of the University's Great Teacher Award, he is the author of *Death and Existence: A Conceptual History of Human Mortality* and *The Silence of God: Meditations on Prayer*. James P. Carse lives in New York City.

FINITE AND INFINITE GAMES

JAMES P. CARSE

PENGUIN BOOKS

Penguin Books Ltd, Harmondsworth, Middlesex, England
Viking Penguin Inc., 40 West 23rd Street, New York, New York 10010, U.S.A.
Penguin Books Australia Ltd, Ringwood, Victoria, Australia
Penguin Books Canada Limited, 2801 John Street, Markham, Ontario, Canada L3R 1B4
Penguin Books (N.Z.) Ltd, 182–190 Wairau Road, Auckland 10, New Zealand

First published in the U.S.A. by the Free Press, a Division of Macmillan, Inc., New York 1986
Published in Penguin Books 1987

Made and printed in Great Britain by
Hazell Watson & Viney Limited,
Member of the BPCC Group,
Aylesbury, Bucks

This book is dedicated to
Alisa, Keene, and Jamie, of course.

CONTENTS

ONE

There Are at Least Two Kinds of Games

1

THERE ARE at least two kinds of games. One could be called finite, the other infinite.

A finite game is played for the purpose of winning, an infinite game for the purpose of continuing the play.

2

If a finite game is to be won by someone it must come to a definitive end. It will come to an end *when* someone has won.

We know that someone has won the game when all the players have agreed who among them is the winner. No other condition than the agreement of the players is absolutely required in determining who has won the game.

It may appear that the approval of the spectators, or the referees, is also required in the determination of the winner. However, it is simply the case that if the players do not agree on a winner, the game has not come to a decisive conclusion— *and* the players have not satisfied the original purpose of playing. Even if they are carried from the field and forcibly blocked from further play, they will not consider the game ended.

Suppose the players all agree, but the spectators and the referees do not. Unless the players can be persuaded that their agreement was mistaken, they will not resume the play—indeed, they *cannot* resume the play. We cannot imagine players returning to the field and truly *playing* if they are convinced the game is over.

There is no finite game unless the players freely choose to play it. No one can play who is forced to play.

It is an invariable principle of all play, finite and infinite, that whoever plays, plays freely. Whoever *must* play, cannot *play*.

3

Just as it is essential for a finite game to have a definitive ending, it must also have a precise beginning. Therefore, we can speak of finite games as having temporal boundaries—to which, of course, all players must agree. But players must agree to the establishment of spatial and numerical boundaries as well. That is, the game must be played within a marked area, and with specified players.

Spatial boundaries are evident in every finite conflict, from the simplest board and court games to world wars. The opponents in World War II agreed not to bomb Heidelberg and Paris and declared Switzerland outside the boundaries of conflict. When unnecessary and excessive damage is inflicted by one of the sides in warfare, a question arises as to the legitimacy of the victory that side may claim, even whether it has been a war at all and not simply gratuitous unwarranted violence. When Sherman burned his way from Atlanta to the sea, he so ignored the sense of spatial limitation that for many persons the war was not legitimately won by the Union Army, and has in fact never been concluded.

Numerical boundaries take many forms but are always applied in finite games. Persons are selected for finite play. It is the case that we cannot play if we must play, but it is also the case *that we cannot play alone.* Thus, in every case, we must find an opponent, and in most cases teammates, who are willing to join in play with us. Not everyone who wishes to do so may play for, or against, the New York Yankees. Neither may they be electricians or agronomists by individual choice, without the approval of their potential colleagues and competitors.

Because finite players cannot select themselves for play, there is never a time when they cannot be removed from the game, or when the other contestants cannot refuse to play with them. The license never belongs to the licensed, nor the commission to the officer.

What is preserved by the constancy of numerical boundaries, of course, is the possibility that all contestants can agree on an eventual winner. Whenever persons may walk on or off the field of play as they wish, there is such a confusion of participants that none can emerge as a clear victor. Who, for example, won the French Revolution?

4

To have such boundaries means that the date, place, and membership of each finite game are *externally* defined. When we say of a particular contest that it began on September 1, 1939, we are speaking from the perspective of world time; that is, from the perspective of what happened before the beginning of the conflict and what would happen after its conclusion. So also with place and membership. A game is played in *that* place, with *those* persons.

The world is elaborately marked by boundaries of contest, its people finely classified as to their eligibilities.

5

Only one person or team can *win* a finite game, but the other contestants may well be *ranked* at the conclusion of play.

Not everyone can be a corporation president, although some who have competed for that prize may be vice presidents or district managers.

There are many games we enter not expecting to win, but in which we nonetheless compete for the highest possible ranking.

6

In one respect, but only one, an infinite game is identical to a finite game: Of infinite players we can also say that if they play they play freely; if they *must* play, they cannot *play*.

Otherwise, infinite and finite play stand in the sharpest possible contrast.

Infinite players cannot say when their game began, nor do they care. They do not care for the reason that their game is not bounded by time. Indeed, the only purpose of the game is to prevent it from coming to an end, to keep everyone in play.

There are no spatial or numerical boundaries to an infinite game. No world is marked with the barriers of infinite play, and there is no question of eligibility since anyone who wishes may play an infinite game.

While finite games are externally defined, infinite games are internally defined. The time of an infinite game is not world time, but time created within the play itself. Since each play of an infinite game eliminates boundaries, it opens to players a new horizon of time.

For this reason it is impossible to say how long an infinite game has been played, or even can be played, since duration can be measured only externally to that which endures. It is also impossible to say in which world an infinite game is played, though there can be any number of worlds within an infinite game.

7

Finite games can be played within an infinite game, but an infinite game cannot be played within a finite game.

Infinite players regard their wins and losses in whatever finite games they play as but moments in continuing play.

8

If finite games must be externally bounded by time, space, and number, they must also have internal limitations on what the players can do to and with each other. To agree on internal limitations is to establish *rules of play*.

The rules will be different for each finite game. It is, in fact, by knowing what the rules are that we know what the game is.

What the rules establish is a range of limitations on the players: each player must, for example, start behind the white line, or have all debts paid by the end of the month, charge patients no 'more than they can reasonably afford, or drive in the right lane.

In the narrowest sense rules are not laws; they do not mandate specific behavior, but only restrain the freedom of the players, allowing considerable room for choice within those restraints.

If these restraints are not observed, the outcome of the game is directly threatened. *The rules of a finite game are the contractual terms by which the players can agree who has won.*

9

The rules must be published prior to play, and the players must agree to them before play begins.

A point of great consequence to all finite play follows from this: *The agreement of the players to the applicable rules constitutes the ultimate validation of those rules.*

Rules are not valid because the Senate passed them, or because heroes once played by them, or because God pronounced them through Moses or Muhammad. They are valid only if and when players freely play by them.

There are no rules that require us to obey rules. If there were, there would have to be a rule for those rules, and so on.

10

If the rules of a finite game are unique to that game it is evident that *the rules may not change in the course of play*—else a different game is being played.

It is on this point that we find the most critical distinction between finite and infinite play: *The rules of an infinite game must change in the course of play.* The rules are changed when the players of an infinite game agree that the play is imperiled by a finite outcome—that is, by the victory of some players and the defeat of others.

The rules of an infinite game are changed to prevent anyone from winning the game and to bring as many persons as possible into the play.

If the rules of a finite game are the contractual terms by which the players can agree who has won, the rules of an infinite game are the contractual terms by which the players agree to continue playing.

For this reason the rules of an infinite game have a different status from those of a finite game. They are like the grammar of a living language, where those of a finite game are like the rules of debate. In the former case we observe rules as a way of continuing discourse with each other; in the latter we observe rules as a way of bringing the speech of another person to an end.

The rules, or grammar, of a living language are always evolving to guarantee the meaningfulness of discourse, while the rules of debate must remain constant.

11

Although the rules of an infinite game may change by agreement at any point in the course of play, it does not follow that any rule will do. It is not in this sense that the game is infinite.

The rules are always designed to deal with specific threats to the continuation of play. Infinite players use the rules to regulate the way they will take the boundaries or limits being forced against their play into the game itself.

The rule-making capacity of infinite players is often challenged by the impingement of powerful boundaries against their play—such as physical exhaustion, or the loss of material resources, or the hostility of nonplayers, or death.

The task is to design rules that will allow the players to continue the game by taking these limits into play—even when death is one of the limits. It is in this sense that the game is infinite.

This is equivalent to saying that no limitation may be imposed against infinite play. Since limits are taken into play, the play itself cannot be limited.

Finite players play within boundaries; infinite players play with boundaries.

12

Although it may be evident enough in theory that whoever plays a finite game plays freely, it is often the case that finite players will be unaware of this absolute freedom and will come to think that whatever they do they *must* do. There are several possible reasons for this:

—We saw that finite players must be selected. While no one is forced to remain a lawyer or a rodeo performer or a kundalini yogi after being selected for these roles, each role is nonetheless surrounded both by ruled restraints and expectations on the part of others. One senses a compulsion to maintain a certain level of performance, because permission to play in these games can be canceled. We cannot do whatever we please and remain lawyers or yogis—and yet we could not be either unless we pleased.

—Since finite games are played *to be won,* players make every move in a game in order to win it. Whatever is not done in the interest of winning is not part of the game. The constant attentiveness of finite players to the progress of the competition can lead them to believe that every move they make they must make.

—It may appear that the prizes for winning are indispensable, that without them life is meaningless, perhaps even impossible. There are, to be sure, games in which the stakes seem to be life and death. In slavery, for example, or severe political oppression, the refusal to play the demanded role may be paid for with terrible suffering or death.

Even in this last, extreme case we must still concede that whoever takes up the commanded role does so by choice. Certainly the price for refusing it is high, but that there is a price at all points to the fact that oppressors themselves acknowledge that even the weakest of their subjects must *agree* to be oppressed. If the subjects were unresisting puppets or automatons, no threat would be necessary, and no price would be paid—thus the satire of the putative ideal of oppressors in Huxley's Gammas, Orwell's Proles, and Rossum's Universal Robots (Capek).

Unlike infinite play, finite play is limited from without; like infinite play, those limitations must be chosen by the player since no one is under any necessity to play a finite game. Fields of play simply do not impose themselves on us. Therefore, all the limitations of finite play are self-limitations.

13

To account for the large gap between the *actual freedom* of finite players to step off the field of play at any time and the *experienced necessity* to stay at the struggle, we can say that as finite players we somehow veil this freedom from ourselves.

Some self-veiling is present in all finite games. Players must intentionally forget the inherently voluntary nature of their play, else all competitive effort will desert them.

From the outset of finite play each part or position must be taken up with a certain seriousness; players must see themselves *as* teacher, *as* light-heavyweight, *as* mother. In the proper exercise of such roles we positively believe we are the persons those roles portray. Even more: we make those roles believable to others. It is in the nature of acting, Shaw said, that we are not to see this woman as Ophelia, but Ophelia as this woman.

If the actress is so skillful that we do see Ophelia as this woman, it follows that we do not see performed emotions and hear recited words, but a person's true feelings and speech. To some extent the actress does not see herself performing but feels her performed emotion and actually says her memorized lines—and yet the very fact that they are performed means that the words and feelings belong to the role and not to the actress. In fact, it is one of the requirements of her craft that she keep her own person distinct from the role. What she feels as the person she is has nothing to do with Ophelia and must not enter into her playing of the part.

Of course, not for a second will this woman in her acting be unaware that she is acting. She never forgets that she has veiled herself sufficiently to play this role, that she has chosen to forget for the moment that she is this woman and not Ophelia. But then, neither do we as audience forget we are audience. Even though we see this woman as Ophelia, we are never in doubt that she is not. We are in complicity with her veil. We allow her performed emotions to affect us, perhaps powerfully. But we never forget that we *allow* them to do so.

So it is with all roles. Only freely can one step into the role of mother. Persons who assume this role, however, must suspend their freedom with a proper seriousness in order to act as the role requires. A mother's words, actions, and feelings belong to the role and not to the person—although some persons may veil themselves so assiduously that they make their performance believable even to themselves, overlooking any distinction between a mother's feelings and their own.

The issue here is not whether self-veiling can be avoided, or even should be avoided. Indeed, no finite play is possible without it. The issue is whether we are ever willing to drop the veil and openly acknowledge, if only to ourselves, that we have freely chosen to face the world through a mask. Consider the actress whose skill at making Ophelia appear as this woman demonstrates the clarity with which she can distinguish the role from herself. Is it not possible that when she leaves the stage she does not give up acting, but simply leaves off one role for another, say the role of "actress," an abstracted personage whose public behavior is carefully scripted and produced? At which point do we confront the fact that we live one life and perform another, or others, attempting to make our momentary forgetting true and lasting forgetting?

What makes this an issue is not the morality of masking ourselves. It is rather that self-veiling is a contradictory act— a free suspension of our freedom. I cannot forget that I have forgotten. I may have used the veil so successfully that I have made my performance believable to myself. I may have convinced myself I am Ophelia. But credibility will never suffice to undo the contradictoriness of self-veiling. "To believe is to know you believe, and to know you believe is not to believe" (Sartre).

If no amount of veiling can conceal the veiling itself, the issue is how far we will go in our seriousness at self-veiling, and how far we will go to have others act in complicity with us.

14

Since finite games can be played within an infinite game, infinite players do not eschew the performed roles of finite play. On the contrary, they enter into finite games with all the appropriate energy and self-veiling, but they do so without the seriousness of finite players. They embrace the abstractness of finite games *as* abstractness, and therefore take them up not seriously, but playfully. (The term "abstract" is used here according to Hegel's familiar definition of it as the substitution of a part of the whole for the whole, the whole being "concrete.") They freely use masks in their social engagements, but not without acknowledging to themselves and others that they are masked. For that reason they regard each participant in finite play as *that person playing* and not *as a role played by someone.*

Seriousness is always related to roles, or abstractions. We are likely to be more serious with police officers when we find them uniformed and performing their mandated roles than when we find them in the process of changing into their uniforms. Seriousness always has to do with an established script, an ordering of affairs completed somewhere outside the range of our influence. We are playful when we engage others at the level of choice, when there is no telling in advance where our relationship with them will come out—when, in fact, no one has an outcome to be imposed on the relationship, apart from the decision to continue it.

To be playful is not to be trivial or frivolous, or to act as though nothing of consequence will happen. On the contrary, when we are playful with each other we relate as free persons, and the relationship is open to surprise; *everything* that happens is of consequence. It is, in fact, seriousness that closes itself to consequence, for seriousness is a dread of the unpredictable outcome of open possibility. To be serious is to press for a specified conclusion. To be playful is to allow for possibility whatever the cost to oneself.

There is, however, a familiar form of playfulness often associated with situations protected from consequence—where no matter what we do (within certain limits), nothing will come of it. This is not playing so much as it is *playing at,* a harmless disregard for social constraints. While this is by no means excluded from infinite play, it is not the same as infinite play.

By relating to others as they move out of their own freedom and not out of the abstract requirements of a role, infinite players are concrete persons engaged with concrete persons. For that reason an infinite game cannot be abstracted, for it is not a part of the whole presenting itself as the whole, but the whole that knows it is the whole. We cannot say a person played this infinite game or that, as though the rules are independent of the concrete circumstances of play. It can be said only that these persons played with each other and in such a way that what they began cannot be finished.

15

Inasmuch as a finite game is intended for conclusion, inasmuch as its roles are scripted and performed for an audience, we shall refer to finite play as *theatrical*. Although script and plot do not seem to be written in advance, we are always able to look back at the path followed to victory and say of the winners that they certainly knew how to act and what to say.

Inasmuch as infinite players avoid any outcome whatsoever, keeping the future open, making all scripts useless, we shall refer to infinite play as *dramatic*.

Dramatically, one chooses to be a mother; theatrically, one takes on the role of mother.

16

One obeys the rules in a finite game in order to play, but playing does not consist only in obeying rules.

The rules of a finite game do not constitute a script. A script is composed according to the rules but is not identical to the rules. The script is the record of the actual exchanges between players—whether acts or words—and therefore cannot be written down beforehand. In all true finite play the scripts are composed in the course of play.

This means that *during the game* all finite play is dramatic, since the outcome is yet unknown. That the outcome is not known is what makes it a true game. The theatricality of finite play has to do with the fact that there is an outcome.

Finite play is dramatic, but only provisionally dramatic. As soon as it is concluded we are able to look backward and see how the sequence of moves, though made freely by the competitors, could have resulted only in this outcome. We can see how every move fit into a sequence that made it inevitable that this player would win.

The fact that a finite game is provisionally dramatic means that it is the intention of each player to eliminate its drama by making a preferred end inevitable. It is the desire of all finite players to be *Master Players,* to be so perfectly skilled in their play that nothing can surprise them, so perfectly trained that every move in the game is foreseen at the beginning. A true Master Player plays as though the game is already in the past, according to a script whose every detail is known prior to the play itself.

17

Surprise is a crucial element in most finite games. If we are not prepared to meet each of the possible moves of an opponent, our chances of losing are most certainly increased.

It is therefore by surprising our opponent that we are most likely to win. Surprise in finite play is the triumph of the past over the future. The Master Player *who already knows* what moves are to be made has a decisive advantage over the unprepared player *who does not yet know* what moves will be made.

A finite player is trained not only to anticipate every future possibility, but to control the future, to *prevent* it from altering the past. This is the finite player in the mode of seriousness with its dread of unpredictable consequence.

Infinite players, on the other hand, continue their play in the expectation of being surprised. If surprise is no longer possible, all play ceases.

Surprise causes finite play to end; it is the reason for infinite play to continue.

Surprise in infinite play is the triumph of the future over the past. Since infinite players do not regard the past as having an outcome, they have no way of knowing what has been begun there. With each surprise, the past reveals a new beginning in itself. Inasmuch as the future is always surprising, the past is always changing.

Because finite players are trained to prevent the future from altering the past, they must hide their future moves. The unprepared opponent must be kept unprepared. Finite players must appear to be something other than what they are. Everything about their appearance must be concealing. To appear is not to appear. All the moves of a finite player must be deceptive: feints, distractions, falsifications, misdirections, mystifications.

Because infinite players prepare themselves to be surprised by the future, they play in complete openness. It is not an openness as in *candor*, but an openness as in *vulnerability*. It is not a matter of exposing one's unchanging identity, the true self that has always been, but a way of exposing one's ceaseless growth, the dynamic self that has yet to be. The infinite player does not expect only to be amused by surprise, but to be transformed by it, for surprise does not alter some abstract past, but one's own personal past.

To be prepared against surprise is to be *trained*. To be prepared for surprise is to be *educated*.

Education discovers an increasing richness in the past, because it sees what is unfinished there. Training regards the past as finished and the future as to be finished. Education leads toward a continuing self-discovery; training leads toward a final self-definition.

Training repeats a completed past in the future. Education continues an unfinished past into the future.

18

What one wins in a finite game is a title.

A title is the acknowledgment of others that one has been the winner of a particular game. Titles are public. They are for others to notice. I expect others to address me according to my titles, but I do not address myself with them—unless, of course, I address myself as an other. The effectiveness of a title depends on its visibility, its noticeability to others.

19

Any given finite game can be played many times, although each occasion of its occurrence is unique. The game that was played at that *time* by *those* players can never be played again.

Since titles are timeless, but exist only so far as they are acknowledged, we must find means to guarantee the memory of them. The birettas of dead cardinals are suspended from the ceilings of cathedrals, as it were forever; the numbers of great athletes are "retired" or withdrawn from all further play; great achievements are carved in imperishable stone or memorialized by perpetual flames.

Some titles are inherited, though only when the bloodline or some other tangible connection with the original winner has been established, suggesting that the winners have continued to exist in their descendants. The heirs to titles are therefore obliged to display the appropriate emblems: a coat of arms or identifiable styles of speech, clothing, or behavior.

It is a principal function of society to validate titles and to assure their perpetual recognition.

20

It is in connection with the timelessness of titles that we can first discern the importance of death to both finite and infinite games *and* the great difference between the ways death is understood in each.

A finite game must always be won with a terminal move, a final act within the boundaries of the game that establishes the winner beyond any possibility of challenge. A terminal move results, in other words, in the death of the opposing player *as player*. The winner kills the opponent. The loser is dead in the sense of being incapable of further play.

Properly speaking, life and death *as such* are rarely the stakes of a finite game. What one wins is a title; and when the loser of a finite game is declared dead to further play, it is equivalent to declaring that person utterly without title—

a person to whom no attention whatsoever need be given. Death, in finite play, is the triumph of the past over the future, a condition in which no surprise is possible.

For this reason, death for a finite player need have nothing to do with the physical demise of the body; it is not a reference to a corporeal state. There are two ways in which death is commonly associated with the fate of the body: One can be dead in life, or one can be alive in death.

Death in life is a mode of existence in which one has ceased all play; there is no further striving for titles. All competitive engagement with others has been abandoned. For some, though not for all, death in life is a misfortune, the resigned acceptance of a loser's status, a refusal to hold any title up for recognition. For others, however, death in life can be regarded as an achievement, the result of a spiritual discipline, say, intended to extinguish all traces of struggle with the world, a liberation from the need for any title whatsoever. "Die before ye die," declare the Sufi mystics.

Life in death concerns those who are titled and whose titles, since they are timeless, may not be extinguished by death. Immortality, in this case, is not a reward but the condition necessary to the possession of rewards. Victors live forever not because their souls are unaffected by death but because their titles must not be forgotten.

It was not merely the souls of the Egyptian pharaohs that passed on into the afterlife, but their complete offices and roles, along with all the tangible reminders of their earthly triumphs—including servants put to death that they might accompany their titled masters into eternity. For Christian saints "death has lost its sting" not because there is something inherently imperishable in the human soul, but because they have fought the good fight, and they have successfully pressed on "toward the goal for the prize of the upward call of God in Christ Jesus" (Paul).

Soldiers commonly achieve a life in death. Soldiers fight not to stay alive but to save the nation. Those who do fight only to protect themselves are, in fact, considered guilty of the highest military crimes. Soldiers who die fighting the enemy, however, receive the nation's highest reward: They are declared unforgettable. Even unknown soldiers are memorialized—though their names have been lost, their titles will not be.

What the winners of finite games achieve is not properly an after*life* but an after*world*, not continuing existence but continuing recognition of their titles.

21

There are games in which the stakes do seem to be life and death.

Extreme forms of bondage sometimes offer persons the privilege of staying alive in exchange for their play—and death for refusing to play. There is, however, something odd in this exchange. A slave does not so much *receive* a life as *give* a life—a life whose only function is to reflect the master's superiority. The slave's life is the property of the master; the slave exists only as an emblem of the master's prior victories.

A slave can have life only by giving it away. "He who loves his life loses it, and he who hates his life in this world will keep it for eternal life" (Jesus).

Perhaps a more common example of such life-or-death forms of bondage is found in those persons who resort to expensive medical strategies to be cured of life-threatening illness. They, too, seem to be giving life away in order to win it back. So also are those who observe special diets or patterns of life designed to prolong their youth and to postpone aging and death indefinitely; they hate their life in this world *now* in order that they may have it *later*. And just as with slaves, the life they receive is given to them by others: doctors, yogis, or their anonymous admirers.

When life is viewed by a finite player as the award to be won, then death is a token of defeat. Death is not, therefore, chosen, but inflicted. It happens to one when the struggle against it fails. Death comes as a judgment, a dishonor, a sign of certain weakness. Death for the finite player is deserved, *earned*. "The wages of sin is death" (Paul).

If the losers are dead, the dead are also losers.

There is a contradiction here: If the *prize* for winning finite play is life, then the players are not properly alive. They are competing *for* life. Life, then, is not play, but the outcome of play. Finite players play to live; they do not live their playing. Life is therefore deserved, bestowed, possessed, won. It is not lived. "Life itself appears only as a *means to life*" (Marx).

This is a contradiction common to all finite play. Because the purpose of a finite game is to bring play to an end with the victory of one of the players, each finite game is played to end itself. The contradiction is precisely that all finite play is play against itself.

22

Death, for finite players, is abstract, not concrete. It is not the whole person, but only an abstracted fragment of the whole, that dies in life or lives in death.

So also is life abstract for finite players. It is not the whole person who lives. If life is a means to life, we must abstract ourselves, but only for the sake of winning an abstraction.

Immortality, therefore, is the triumph of such abstraction. It is a *state of unrelieved theatricality*. An immortal soul is a person who cannot help but continue living out a role already scripted. An immortal person could not choose to die nor, for the same reason, choose to live. Immortality is serious and in no way playful. One's actions can have no consequence beyond themselves. There are no surprises in the afterworld.

Of course, immortality of the *soul*—the bare soul, cleansed of any personality traces—is rarely what is desired in the yearning for immortality. "The information that my soul is to last forever could then be of no more personal concern to me than the news that my appendix is to be preserved eternally in a bottle" (Flew). More often what one intends to preserve is a public personage, a permanently veiled selfhood. Immortality is the state of forgetting that we have forgotten—that is, overlooking the fact that we freely decided to enter into finite play, a decision in itself playful and not serious.

Immortality is therefore the supreme example of the contradictoriness of finite play: It is a life one cannot live.

23

Infinite players die. Since the boundaries of death are always part of the play, the infinite player does not die at the end of play, but in the course of play.

The death of an infinite player is dramatic. It does not mean that the game comes to an end with death; on the contrary, infinite players offer their death as a way of continuing the play. For that reason they do not play for their own life; they live for their own play. But since that play is always with others, it is evident that infinite players both live and die for the continuing life of others.

Where the finite player plays for immortality, the infinite player plays as a mortal. In infinite play one chooses to be mortal inasmuch as one always plays dramatically, that is, toward the open, toward the horizon, toward surprise, where nothing can be scripted. It is a kind of play that requires complete vulnerability. To the degree that one is protected against the future, one has established a boundary and no longer plays with but against others.

Death is a defeat in finite play. It is inflicted when one's boundaries give way and one falls to an opponent. The finite player dies under the terminal move of another.

Although infinite players choose mortality, they may not know when death comes, but we can always say of them that "they die at the right time" (Nietzsche).

The finite play for life is serious; the infinite play of life is joyous. Infinite play resounds throughout with a kind of laughter. It is not a laughter at others who have come to an unexpected end, having thought they were going somewhere else. It is laughter *with* others with whom we have discovered that the end we thought we were coming to has unexpectedly opened. We laugh not at what has surprisingly come to be impossible for others, but over what has surprisingly come to be possible with others.

24

Infinite play is inherently *paradoxical*, just as finite play is inherently *contradictory*. Because it is the purpose of infinite players to continue the play, they do not play for themselves.

The contradiction of finite play is that the players desire to *bring play to an end for themselves*. The paradox of infinite play is that the players desire to *continue the play in others*. The paradox is precisely that they play only when others go on with the game.

Infinite players play best when they become least necessary to the continuation of play. It is for this reason they play as mortals.

The joyfulness of infinite play, its laughter, lies in learning to start something we cannot finish.

25

If finite players acquire *titles* from winning their games, we must say of infinite players that they have nothing but their *names*.

Names, like titles, are given. Persons cannot name themselves any more than they can entitle themselves. However, unlike titles, which are given for what a person has done, a name is given at birth—at a time when a person cannot yet have done anything. Titles are given at the end of play, names at the beginning.

When a person is known by title, the attention is on a completed past, on a game already concluded, and not therefore to be played again. A title effectively takes a person out of play.

When a person is known only by name, the attention of others is on an open future. We simply cannot know what to expect. Whenever we address each other by name we ignore all scripts, and open the possibility that our relationship will become deeply reciprocal. That I cannot now predict your future is exactly what makes mine unpredictable. Our futures enter into each other. What is your future, and mine, becomes ours. We prepare each other for surprise.

Titles are abstractions; names are always concrete.

It can happen that when persons are distinctly identified as winners their names can have the force of titles. We sometimes act "to clear our name" of aspersions, or to defend the "good name of our family." Names can even become titles in the formal sense, such as "Caesar," or "Napoleon," or "the name of Jesus which is above every name" (Paul). When Jesus is regarded by way of a title instead of a name, he becomes an abstracted, theatrical role, a person with whom we can share no future, rather a Master Player in whose future we live in a manner that has already been scripted, or decided, for us. "Before Abraham was, I am," Jesus said of himself in the Gospel of John.

26

Titles, then, point backward in time. They have their origin in an unrepeatable past.

Titles are theatrical. Each title has a specified ceremonial form of address and behavior. Titles such as Captain, Mrs., Lord, Esquire, Professor, Comrade, Father, Under Secretary, signal not only a *mode of address* with its appropriate deference or respect, but also a *content of address* (only certain subjects are suitable for discussion with the Admiral of the Fleet or the District Attorney or the Holy Mother), and a *manner of address* (shaking hands, kneeling, prostrating or crossing oneself, saluting, bowing, averting the eyes, or standing in silence).

The mode and content of address and the manner of behavior are recognitions of the areas in which titled persons *are no longer in competition.* There are precise ways in which one may no longer compete with the Dalai Lama or the Heavyweight Champion of the World. There is no possible action by which one may deprive them of their titles to contests now in the past. Therefore, insofar as we recognize their titles we withdraw from any contest with them in those areas.

27

The titled are *powerful.* Those around them are expected to yield, to withdraw their opposition, and to conform to their will—in the arena in which the title was won.

The exercise of power always presupposes resistance. Power is never evident until two or more elements are in opposition. Whichever element can move another is the more powerful. If no one else ever strove to be a Boddhisattva or the Baton Twirling Champion of the State of Indiana, those titles would be powerless—no one would defer to them.

The exercise of power also presupposes a closed field and finite units of time. My power is determined by the amount of resistance I can displace *within given spatial and temporal limits.* The question is not whether I can lift ten pounds, but whether I can lift ten pounds five feet off the ground in one second—or within some other precise limitation of time and space. The establishment of the limits makes it possible to know how powerful I am *in relation to others.*

Power is always measured in units of comparison. In fact, it is a term of competition: How much resistance can I overcome relative to others?

Power is a concept that belongs only in finite play. But power is not properly measurable until the game is completed—until the designated period of time has run out. During the course of play we cannot yet determine the power of the players, because to the degree that it is genuine play the outcome is unknown. A player who is being pushed all over the field by an apparently superior opponent may display an unsuspected burst of activity at the end and take the victory. Until the final hours of the count in the presidential election of 1948 many Americans thought that Harry Truman was a far weaker candidate than Thomas Dewey.

To speak meaningfully of a person's power is to speak of what that person *has already completed* in one or another closed field. To see power is to look backward in time.

Inasmuch as power is determined by the outcome of a game, one does not win *by being powerful;* one wins *to be powerful.* If one has sufficient power to win before the game has begun, what follows is not a game at all.

One can be powerful only through the possession of an acknowledged title—that is, only by the ceremonial deference of others. Power is never one's own, and in that respect it shows the contradiction inherent in all finite play. I can be powerful only by not playing, by showing that the game is over. I can therefore *have* only what powers others *give* me. Power is bestowed by an audience after the play is complete.

Power is contradictory, and theatrical.

28

It may seem implausible to claim that power is a matter of deference to titles. If anything appears to be a permanent feature of reality it is power—the constant impingement on us of superior forces both without and within. Everything from changes in the weather and acts of national governments to the irresistible push of instinct and the process of aging seems to confirm us as helpless creatures of circumstance—and to that degree powerless. It seems plainly false to say that power is theatrical.

And yet, the theatrical nature of power seems to be consistent with the principle arrived at earlier: Whoever *must* play cannot *play*. The intuitive idea in that principle is that no one can engage us competitively unless we fully cooperate, unless we join the game and join it to win. Because power is measurable only in comparative—that is, competitive—terms, it presupposes some kind of cooperation. If we defer to titled winners, it is only because we regard ourselves as losers. To do so is freely to take part in the theater of power.

There certainly are acts of government, or acts of nature, or acts of god that far exceed any contravening ability of our own, but it is unlikely that we would consider ourselves losers in relation to them. We are not *defeated* by floods or genetic disease or the rate of inflation. It is true that these are *real*, but we do not play *against reality*; we play *according to reality*. We do not eliminate weather or genetic influence but accept them as the realities that establish the context of play, the limits within which we are to play.

If I accept death as inevitable, I do not struggle against mortality. I struggle as a mortal.

All the limitations of finite play are self-limitations.

29

Power is a feature only of finite games. It is not dramatic but theatrical. How then do infinite players contend with power? Infinite play is always dramatic; its outcome is endlessly open. There is no way of looking back to make a definitive assessment of the power or weakness of earlier play. Infinite players look forward, not to a victory in which the past will achieve a timeless meaning, but toward ongoing play in which the past will require constant reinterpretation. Infinite players do not *oppose* the actions of others, but *initiate* actions of their own in such a way that others will respond by initiating *their* own.

We need a term that will stand in contrast to "power" as it acquires its meaning in finite play. Let us say that where the finite player plays *to be powerful* the infinite player plays *with strength.*

A powerful person is one who brings the past to an outcome, settling all its unresolved issues. A strong person is one who carries the past into the future, showing that none of its issues is capable of resolution. Power is concerned with what has already happened; strength with what has yet to happen. Power is finite in amount. Strength cannot be measured, because it is an opening and not a closing act. Power refers to the freedom persons have within limits, strength to the freedom persons have with limits.

Power will always be restricted to a relatively small number of selected persons. Anyone can be strong.

Strength is paradoxical. I am not strong because I can force others to do what I wish *as a result of my play with them,* but because I can allow them to do what they wish *in the course of my play with them.*

30

Although anyone who wishes can be an infinite player, and although anyone can be strong, we are not to suppose that power cannot work irremediable damage on infinite play. Infinite play cannot prevent or eliminate evil. Though infinite players are strong, they are not powerful and do not attempt to become powerful.

Evil is the termination of infinite play. It is infinite play coming to an end in *unheard silence.*

Unheard silence does not necessarily mean the death of the player. Unheard silence is not the loss of the player's voice, but the loss of listeners for that voice. It is an evil when the drama of a life does not continue in others for reason of their deafness or ignorance.

There are silences that can be heard, even from the dead and from the severely oppressed. Much is recoverable from an apparently forgotten past. Sensitive and faithful historians can learn much of what has been lost, and much therefore that can be continued.

There are silences, however, that will never and can never be heard. There is much evil that remains beyond redemption. When Europeans first landed on the North American continent the native population spoke as many as ten thousand distinct languages, each with its own poetry and treasury of histories and myths, its own ways of living in harmony with the spontaneities of the natural environment. All but a very few of those tongues have been silenced, their cultures forever lost to those of us who stand ignorantly in their place.

Evil is *not* the termination of a finite game. Finite players, even those who play for their own lives, know the stakes of the games they freely choose to play.

Evil is not the attempt to eliminate the play of another according to published and accepted rules, but to eliminate the play of another regardless of the rules. Evil is not the acquisition of power, but the expression of power. It is the forced recognition of a title—and therein lies the contradiction of evil, for recognition cannot be forced. The Nazis did not compete with the Jews for a title, but demanded recognition of a title without competition. This could be achieved, however, only by silencing the Jews, only by hearing nothing from them. They were to die in silence, along with their culture, without anyone noticing, not even those who managed the institutions and instruments of death.

31

Evil is never intended as evil. Indeed, the contradiction inherent in all evil is that it originates in the desire to eliminate evil. "The only good Indian is a dead Indian."

Evil arises in the honored belief that history can be tidied up, brought to a sensible conclusion. It is evil to act as though the past is bringing us to a specifiable end. It is evil to assume that the past will make sense only if we bring it to an issue we have clearly in view. It is evil for a nation to believe it is "the last, best hope on earth." It is evil to think history is to end with a return to Zion, or with the classless society, or with the Islamicization of all living infidels.

Your history does not belong to me. We live with each other in a common history.

Infinite players understand the inescapable likelihood of evil. They therefore do not attempt to eliminate evil in others, for to do so is the very impulse of evil itself, and therefore a contradiction. They only attempt paradoxically to recognize in themselves the evil that takes the form of attempting to eliminate evil elsewhere.

Evil is not the inclusion of finite games in an infinite game, but the restriction of all play to one or another finite game.

TWO

No One Can Play a Game Alone

32

No ONE CAN PLAY a game alone. One cannot be human by oneself. There is no selfhood where there is no community. We do not relate to others as the persons we are; we are who we are in relating to others.

Simultaneously the others with whom we are in relation are themselves in relation. We cannot relate to anyone who is not also relating to us. Our social existence has, therefore, an inescapably fluid character. This is not to say that we live in a fluid context, but that our lives are themselves fluid. As in the Zen image we are not the stones over which the stream of the world flows; we are the stream itself. As we shall see, this ceaseless change does not mean discontinuity; rather change is itself the very basis of our continuity as persons. Only that which can change can continue: this is the principle by which infinite players live.

The fluidity of our social and therefore personal existence is a function of our essential freedom—the kind of freedom indicated in the formula "Who must play, cannot play." Of course, as we have seen, finite games cannot have fluid boundaries, for if they do it will be impossible to agree on winners. But finite games float, as it were, in the unconstrained choice each player makes in entering and continuing the play. Finite games sometimes appear, therefore, to have fixed points of social reference. Not only are there true and false ways of loving your country, for example; there is a positive requirement that you do so.

It is this essential fluidity of our humanness that is irreconcilable with the seriousness of finite play. It is, therefore, this fluidity that presents us with an unavoidable challenge: how to contain the serious within the truly playful; that is, how to keep all our finite games in infinite play.

This challenge is commonly misunderstood as the need to find room for playfulness within finite games. This is what was referred to above as playing at, or perhaps playing around, a kind of play that has no consequence. This is the sort of playfulness implied in the ordinary sense of such terms as entertainment, amusement, diversion, comic relief, recreation, relaxation. Inevitably, however, seriousness will creep back into this kind of play. The executive's vacation, like the football team's time out, comes to be a device for refreshing the contestant for a higher level of competition. Even the open playfulness of children is exploited through organized athletic, artistic, and educational regimens as a means of preparing the young for serious adult competition.

33

When Bismarck described politics as the art of the possible, he meant, of course, that the possible is to be found somewhere within fixed limits, within social realities. He plainly did not mean that the possible extended to those limits themselves. Such a politics is therefore seriousness itself, especially since politicians of nearly every ideology represent themselves as champions of freedom, doing what is necessary and even distasteful toward the end of enlarging the range of the possible. "We must learn the fine arts of war and independence so that our children can learn architecture and engineering so that their children may learn the fine arts and painting" (John Quincy Adams).

The interest of infinite players has little in common with such politics, since they are not concerned to find how much freedom is available within the given realities—for this is freedom only in the trivial sense of playing at—but are concerned to show how freely we have decided to place these particular boundaries around our finite play. They remind us that political realities do not precede, but follow from, the essential fluidity of our humanness.

This does not mean that infinite players are politically disengaged; it means rather that they are political without having a politics, a paradoxical position easily misinterpreted. To have a politics is to have a set of rules by which one attempts to reach a desired end; to be political—in the sense meant here—is to recast rules in the attempt to eliminate all societal ends, that is, to maintain the essential fluidity of human association.

To be political in the mode of infinite play is by no means to disregard the appalling conditions under which many human beings live, the elimination of which is the professed end of much politics. We can imagine infinite players nodding thoughtfully at Rousseau's famous declaration: "Man is born free, and everywhere he is in chains." They can see that the dream of freedom is universal, that wars are fought to win it, heroes die to protect it, and songs are written to commemorate its attainment. But in the infinite player's vision of political affairs the element of intentionality and willfulness, so easily obscured in the exigencies of public crisis, stands out in clear relief. Therefore, even warfare and heroism are seen with their self-contradictions in full display. No nation can go to war until it has found another that can agree to the terms of the conflict. Each side must therefore be in complicity with the other: Before I can have an enemy, I must persuade another to recognize me as an enemy. I cannot be a hero unless I can first find someone who will threaten my life— or, better, take my life. Once under way, warfare and acts of heroism have all the appearance of necessity, but that appearance is but a veil over the often complicated maneuvers by which the antagonists have arranged their conflict with each other.

Therefore, for infinite players, politics is a form of theatricality. It is the performance of roles before an audience, according to a script whose last scene is known in advance by the performers. The United States did not, for example, lose its war in Southeast Asia so much as lose its audience for a war. No doubt much of the disillusion and bitterness of its warriors comes from the missing final scene—the hero's homecoming to parades or ceremonial burial—an anticipated scene that carries many into battle.

It is because of the essential theatricality of politics that infinite players do not take sides in political issues—at least not seriously. Instead they enter into social conflict dramatically, attempting to offer a vision of continuity and open-endedness in place of the heroic final scene. In doing so they must at the very least draw the attention of other political participants not to what they feel they must do, but to why they feel they must do it.

In their own political engagements infinite players make a distinction between society and culture. *Society* they understand as the sum of those relations that are under some form of public constraint, *culture* as whatever we do with each other by undirected choice. If society is all that a people feels it must do, culture "is the realm of the variable, free, not necessarily universal, of all that cannot lay claim to compulsive authority" (Burckhardt).

The infinite player's understanding of society is not to be confused with, say, natural instinct, or any other form of involuntary activity. Society remains entirely within our free choice in quite the same way that finite competition, however strenuous or costly to the player, never prevents the player from walking off the field of play. Society applies only to those areas of action which are believed to be necessary.

Just as infinite play cannot be contained within finite play, culture cannot be authentic if held within the boundaries of a society. Of course, it is often the strategy of a society to initiate and embrace a culture as exclusively its own. Culture so bounded may even be so lavishly subsidized and encouraged by society that it has the appearance of open-ended activity, but in fact it is designed to serve societal interests in every case—like the socialist realism of Soviet art.

Society and culture are therefore not true opponents of each other. Rather society is a species of culture that persists in contradicting itself, a freely organized attempt to conceal the freedom of the organizers and the organized, an attempt to forget that we have willfully forgotten our decision to enter this or that contest and to continue in it.

34

If we think of society as all that a people does under the veil of necessity, we must also think of it as a single finite game that includes any number of smaller games within its boundaries.

A large society will consist of a wide variety of games—though all somehow connected, inasmuch as they have a bearing on a final societal ranking. Schools are a species of finite play to the degree that they bestow ranked awards on those who win degrees from them. Those awards in turn qualify graduates for competition in still higher games—certain prestigious colleges, for example, and then certain professional schools beyond that, with a continuing sequence of higher games in each of the professions, and so forth. It is not uncommon for families to think of themselves as a competitive unit in a broader finite game for which they are training their members in the struggle for societally visible titles.

Like a finite game, a society is numerically, spatially, and temporally limited. Its citizenship is precisely defined, its boundaries are inviolable, and its past is enshrined.

The power of citizens in a society is determined by their ranking in games that have been played. A society preserves its memory of past winners. Its record-keeping functions are crucial to societal order. Large bureaucracies grow out of the need to verify the numerous entitlements of the citizens of that society.

The power of a society is determined by its victory over other societies in still larger finite games. Its most treasured memories are those of the heroes fallen in victorious battles with other societies. Heroes of lost battles are almost never memorialized. Foch has his monument, but not Petain; Lincoln, but not Jefferson Davis; Lenin, but not Trotsky.

The power in a society is guaranteed and enhanced by the power of a society.

The prizes won by its citizens can be protected only if the society as a whole remains powerful in relation to other societies. Those who desire the permanence of their prizes will work to sustain the permanence of the whole. Patriotism in one or several of its many forms (chauvinism, racism, sexism, nationalism, regionalism) is an ingredient in all societal play.

Because power is inherently patriotic, it is characteristic of finite players to seek a growth of power in a society as a way of increasing the power of a society. It is in the interest of a society therefore to encourage competition within itself, to establish the largest possible number of prizes, for the holders of prizes will be those most likely to defend the society as a whole against its competitors.

35

Culture, on the other hand, is an infinite game. Culture has no boundaries. Anyone can be a participant in a culture—anywhere and at any time.

Because a society maintains careful temporal limits, it understands its past as destiny; that is, its course of history lies between a definitive beginning (the founders of a society are always especially memorialized) and a definitive ending. (The nature of its victory is repeatedly anticipated in official declarations; "to each according to their need, from each according to their ability," for example.)

Because culture as such can have no temporal limits, a culture understands its past not as destiny, but as history, that is, as a narrative that has begun but points always toward the endlessly open. Culture is an enterprise of mortals, disdaining to protect themselves against surprise. Living in the strength of their vision, they eschew power and make joyous play of boundaries.

Society is a manifestation of power. It is theatrical, having an established script. Deviations from the script are evident at once. Deviation is antisocietal and therefore forbidden by society under a variety of sanctions. It is easy to see why deviancy is to be resisted. If persons did not adhere to the standing rules of the society, any number of rules would change, and some would be dropped altogether. This would mean that past winners no longer warrant ceremonial recognition of their titles and are therefore without power—like Russian princes after the Revolution.

It is a highly valued function of society to prevent changes in the rules of the many games it embraces. Such procedures as academic accreditation, licensure of trades and professions, synodical ordination, parliamentary confirmation of official appointments, and the inauguration of political leaders are acts of the larger society allowing persons to compete in the finite games within it.

Deviancy, however, is the very essence of culture. Whoever merely follows the script, merely repeating the past, is culturally impoverished.

There are variations in the quality of deviation; not all divergence from the past is culturally significant. Any attempt to vary from the past in such a way as to cut the past off, causing it to be forgotten, has little cultural importance. Greater significance attaches to those variations that bring the tradition into view in a new way, allowing the familiar to be seen as unfamiliar, as requiring a new appraisal of all that we have been—and therefore of all that we are.

Cultural deviation does not return us to the past, but continues what was begun and not finished in the past. Societal convention, on the other hand, requires that a completed past be repeated in the future. Society has all the seriousness of immortal necessity; culture resounds with the laughter of unexpected possibility. Society is abstract, culture concrete.

Finite games can be played again; they can be played an indefinite number of times. It is true that the winners of a game are always the winners of a game played at that particular time, but the validity of their titles depends on the repeatability of that game. We memorialize early football greats but would not do so if football had vanished after its first decade.

As we have seen, because an infinite game cannot be brought to an end, it cannot be repeated. Unrepeatability is a characteristic of culture everywhere. Mozart's Jupiter Symphony cannot be composed again, nor could Rembrandt's self-portraits be painted twice. Society preserves these works as the prizes of those who have triumphed in their respective games. Culture, however, does not consider the works as the outcome of a struggle, but as moments in an ongoing struggle—the very struggle that culture is. Culture continues what Mozart and Rembrandt had themselves continued by way of their work: an original, or deviant, shaping of the tradition they received, original enough that it does not invite duplication of itself by others, but invites the originality of others in response.

Just as an infinite game has rules, a culture has a tradition. Since the rules of play in an infinite game are freely agreed to and freely altered, a cultural tradition is both adopted and transformed in its adoption.

Properly speaking, a culture does not have a tradition; it is a tradition.

36

It is essential to the identity of a society to forget that it has forgotten that society is always a species of culture. Its citizens must find ways of persuading themselves that their own particular boundaries have been imposed on them, and were not freely chosen by them. For example, it is one thing for persons to choose to be Americans, quite another for persons to choose to be America. Societal thinking easily permits the former, never the latter.

One of the most effective means of self-persuasion available to a citizenry is the bestowal of property. Who actually owns a society's property, and how it is distributed, are far less important than the fact that property exists at all. To understand the peculiar dynamic of property we must return to one of the features of finite play.

What the winner of a finite game wins is a title. A title is the acknowledgment of others that one has been the winner of a particular game. I cannot entitle myself. Titles are theatrical, requiring an audience to bestow and respect them. Power attaches to titles inasmuch as those who acknowledge them accept the fact that the struggle in which the titles were won cannot be taken up again. Possession of the title signifies an agreement that competition is forever closed in that particular game.

It is therefore essential to the effectiveness of every title that it be visible and that in its visibility it point back at the contest in which it was won. The purpose of property is to make our titles visible. Property is emblematic. It recalls to others those areas in which our victories are beyond challenge.

Property may be stolen, but the thief does not therefore own it. Ownership can never be stolen. Titles are timeless, and so is the ownership of property. Nations will sometimes go to war over claims to the ownership of land that go back many centuries. Titles can be inherited, and when they are there is an appropriate transfer of property to the heir—who must, of course, possess the very worthiness by which the inheritor originally secured the title. (An inheritance can often be legally challenged by demonstrating the heir's incompetence or immorality.)

A thief, however, does not mean to steal the title. A thief does not want to take what belongs to someone else. The thief does not compete with me for the articles I have title to, but for the title to those articles. The thief means to win the title, believing that those things to which I claim title belong to no one and are there for the taking. "If you don't take pocket-handkechers and watches," explains the Artful Dodger to Oliver, "some other cove will; so that the coves that lost 'em will be all the worse, and you'll be all the worse too and nobody half a ha'p'orth the better, except the chaps wot gets them—and you've just as good a right to them as they have."

37

One reason for the necessity of a society is its role in ascribing and validating the titles to property. "The great and chief end therefore, of Mens uniting into Commonwealths, and putting themselves under Government, is the preservation of their Property; to which in the state of nature there are many things wanting" (Locke).

When we ask precisely how a society will go about preserving its citizens' property, we can expect the reply that it will do so by the use of force. This reply introduces a dilemma. While it is true that there are ways of forcibly restraining a thief, it is also true that no amount of force can lead the Artful Dodger truly to acknowledge a gentleman's title to the handkerchief in his pocket. Until the young ruffian is persuaded freely to respect that title, he will remain a thief. By extension, this observation applies to the society as a whole. There is no effective pattern of entitlement in a society short of the free agreement of all opponents that the titles to property are in the hands of the actual winners.

No force will establish this agreement. Indeed, the opposite is the case: It is agreement that establishes force. Only those who consent to a society's constraints see them as constraints— that is, as guides to action and not as actions to be opposed.

Those who challenge the existing pattern of entitlements in a society do not consider the designated officers of enforcement powerful; they consider them opponents in a struggle that will determine by its outcome who is powerful. One does not win by power; one wins to be powerful.

Only by free self-concealment can persons believe they obey the law because the law is powerful; in fact, the law is powerful for persons only because they obey it. We do not proceed through a traffic intersection *because* the signal changes, but *when* the signal changes.

This means that a peculiar burden falls on property owners. Since the laws protecting their property will be effective only when they are able to persuade others to obey those laws, they must introduce a theatricality into their ownership sufficiently engaging that their opponents will live by its script.

38

The theatricality of property has, in fact, an elaborate structure that property owners must be at considerable labor to sustain. If property is to be persuasively emblematic, that is, if it is to draw attention to the owner's titles in past victories, a double burden falls on its owners:

First, they must show that the amount of their property corresponds to the difficulty they were under in winning title to it. *Property must be seen as compensation.*

Second, they must show that the type of their property corresponds to the nature of the competition by which title to it was won. *Property must be seen to be consumed.*

Property is appropriately compensatory whenever owners can show that what is gained is no more than what was expended in the effort to acquire it. There must be an equivalency between what the owners have given of themselves and what they have received from others by way of their titles.

Whoever is unable to show a correspondence between wealth and the risks undergone to acquire it, or the talents spent in its acquisition, will soon face a challenge over entitlement. The rich are regularly subject to theft, to taxation, to the expectation that their wealth be shared, as though what they have is not true compensation and therefore not completely theirs.

To be fully compensated for what one gave of oneself in the struggle for a title is to be restored to the condition one was in prior to competition.

Property is an attempt to recover the past. It returns one to precompetitive status. One is compensated for the amount of time spent (and thus lost) in competition.

This attempt to recover the past is, however, a theatrical attempt which can succeed only to the degree that it is conspicuous to its audience. Property must take up space. It must be somewhere—and somewhere obvious. That is, it must exist in such a form that others will come upon it and take notice of it. Our property must intrude on another, stand in another's way, causing one to contend with it. Propertied persons typically have large estates and freedom of movement through the society. At the same time, the property of the rich has the effect of crowding and confining the less propertied. The very poor are typically restricted to narrow geographical limits and are regarded as aliens outside them.

What is at stake here for owners is not the amount of

property as such, but its ability to draw an audience for whom it will be appropriately emblematc; that is, an audience who will see it as just compensation for the effort and skill used in acquiring it.

40

There is a second theatrical requirement that falls on the owners of property. Once they have drawn attention to what they have lost in acquiring what they own, they must then consume what they have gained in a way that recovers the loss. The intuitive principle here is that we cannot be justified in owning what we do not need to use or plan to use. One does not earn money merely to store it away where it will be protected from all possible future use.

Consumption is to be understood as an intentional activity. One does not consume property simply by destroying it—else burning our earned money would suffice—but by using it up in a certain way.

Consumption is a kind of activity that is directly opposite to the very form of engagement by which the title was won. It must be the kind of activity that can convince all observers that the possesser's title to it is no longer in question.

The more powerful we consider persons to be, the less we expect them to do, for their power can come only from that which they have done. After athletic contests in which major titles have been at stake, it is common for the audience to lift the winners to their shoulders, marching them about as if they were helpless—in the sharpest possible contrast to the physical skill and energy they have just displayed. Monarchs and divinities are often borne on ceremonial transports; the very wealthy are driven in carriages or limousines.

Consumption is an activity so different from gainful labor that it shows itself in the mode of leisure, even indolence. We display the success of what we have done by not having to do anything. The more we use up, therefore, the more we show ourselves to be winners of past contests. "Conspicuous abstention from labour therefore becomes the conventional mark of superior pecuniary achievement and the conventional index of reputability; and conversely, since application to productive labour is a mark of poverty and subjection, it becomes inconsistent with a reputable standing in the community" (Veblen).

Just as compensation makes itself conspicuous by taking up space, consumption draws attention to itself by the length of time it continues. Property must not only intrude on others, it must *continue* to intrude on others. The amount of property we have is measurable according to the length of time we remain conspicuous, requiring others to adjust their freedom of movement to our spatial dimensions. It is the common goal of the rich to establish a mode of visibility that will extend itself over generations by executing wills that prevent the rapid exhaustion of their fortune, by endowing societally important institutions, by erecting great buildings in their name. Persons of small victories, of lower rank, do not have property of great temporal value; what they have will be exhausted quickly. Those persons whose victories the society wishes never to forget are given prominent and eternal monuments at the heart of its capital cities, often taking up considerable space, diverting traffic, and standing in the path of casual strollers.

It is apparent to infinite players that wealth is not so much possessed as it is performed.

41

If one of the reasons for uniting into commonwealths is the protection of property, and if property is to be protected less by power as such than by theater, then societies become acutely dependent on their artists—what Plato called *poietai:* the storytellers, the inventors, sculptors, poets, any original thinkers whatsoever.

It is certainly the case that no gentleman will find the Artful Dodger's hand in his pocket so long as it is in the forceful grip of an officer of the law. But any policy of forceful restraint so extreme that it requires an officer for each potential criminal is a formula for quick descent into social chaos.

Some societies develop the belief that they can eliminate thievery by guaranteeing all their members, including thieves, a certain amount of property—the impulse behind much social welfare legislation. But putting a coin into the pocket of the Artful Dodger will hardly convince him that he is no longer a legitimate contender for the coin in mine.

The more effective policy for a society is to find ways of persuading its thieves to abandon their role as competitors for property for the sake of becoming audience to the theater of wealth. It is for this reason that societies fall back on the skill of those poietai who can theatricalize the property relations, and indeed, all the inner structures of each society.

Societal theorists of any subtlety whatever know that such theatricalization must be taken with great seriousness. Without it there is no culture at all, and a society without culture would be too drab and lifeless to be endured. What would

Nazism have been without its musicians, graphic artists, and set designers, without its Albert Speer and Leni Riefenstahl? Even the rigid authoritarian shell of Plato's Republic will be "filled with a multitude of things which are no longer necessities, as for example all kinds of hunters and artists, many of them concerned with shapes and colors, many with music; poets and their auxiliaries, actors, choral dancers, and contractors; and makers of all kinds of instruments, including those needed for the beautification of women" (Plato).

If wealth and might are to be performed, great wealth and great might must be performed brilliantly.

42

While societal thinkers may not overlook the importance of *poiesis,* or creative activity, neither may they underestimate its danger, for the poietai are the ones most likely to remember what has been forgotten—that society is a species of culture.

Societies commonly treat their poietai with considerable ambivalence. The governing bodies of the Soviet Union do not believe that all genuine art must conform to the standards of socialist realism, but they do believe it is always possible to find true art that is compatible with socialist realism; therefore, those artists whose works do not conform to that line may be punished without affecting the integrity of art as such. Plato did not expect his artists to compromise their art, but he did say that there must be "general lines which the poietai must follow in their stories. These lines they will not be able to cross."

The deepest and most consequent struggle of each society is therefore not with other societies, but with the culture that exists within itself—the culture that is itself. Conflict with other societies is, in fact, an effective way for a society to restrain its own culture. Powerful societies do not silence their poietai in order that they may go to war; they go to war as a way of silencing their poietai. Original thinkers can be suppressed through execution and exile, or they can be encouraged through subsidy and flattery to praise the society's heroes. Alexander and Napoleon took their poets and their scholars into battle with them, saving themselves the nuisance of repression and along the way drawing ever larger audiences to their triumph.

Another successful defense of society against the culture within itself is to give artists a place by regarding them as the producers of property, thus elevating the value of consuming art, or owning it. It is notable that very large collections of art, and all the world's major museums, are the work of the very rich or of societies during strongly nationalistic periods. All the principal museums in New York, for example, are associated with the names of the famously rich: Carnegie, Frick, Rockefeller, Guggenheim, Whitney, Morgan, Lehman.

Such museums are not designed to protect the art from people, but to protect the people from art.

43

Culture is likely to break out in a society not when its poietai begin to voice a line contrary to that of the society, but when they begin to ignore all lines whatsoever and concern themselves with bringing the audience back into play—not competitive play, but play that affirms itself as play.

What confounds a society is not serious opposition, but the lack of seriousness altogether. Generals can more easily suffer attempts to oppose their warfare with poiesis than attempts to show warfare as poiesis.

Art that is used against a society or its policies gives up its character as infinite play, and aims for an end. Such art is no less propaganda than that which praises its heroes with high seriousness. Once warfare, or any other societal activity, has been taken into the infinite play of poiesis so that it appears to be either comical or pointless (in the way that, say, beauty is pointless), there is an acute danger that the soldiers will find no audience for their prizes, and therefore no reason to fight for them.

44

Since culture is itself a poiesis, all of its participants are poietai—inventors, makers, artists, storytellers, mythologists. They are not, however, makers of actualities, but makers of possibilities. The creativity of culture has no outcome, no conclusion. It does not result in art works, artifacts, products. Creativity is a continuity that engenders itself in others. "Artists do not create objects, but create by way of objects" (Rank).

Art is not art, therefore, except as it leads to an engendering creativity in its beholders. Whoever takes possession of the objects of art has not taken possession of the art.

Since art is never possession, and always possibility, nothing possessed can have the status of art. If art cannot become property, property is never art—as property. Property draws attention to titles, points backward toward a finished time. Art is dramatic, opening always forward, beginning something that cannot be finished.

Because it is not conclusive, but engendering, culture has

no established catalogue of acceptable activities. We are not artists by reason of having mastered certain skills or exercising specified techniques. Art has no scripted roles for its performers. It is precisely because it has none that it is art. Artistry can be found anywhere; indeed, it can only be *found* anywhere. One must be surprised by it. It cannot be looked for. We do not watch artists to see what they do, but watch what persons do and discover the artistry in it.

Artists cannot be trained. One does not become an artist by acquiring certain skills or techniques, though one can use any number of skills and techniques in artistic activity. The creative is found in anyone who is prepared for surprise. Such a person cannot go to school to be an artist, but can only go to school as an artist.

Therefore, poets do not "fit" into society, not because a place is denied them but because they do not take their "places" seriously. They openly see its roles as theatrical, its styles as poses, its clothing costumes, its rules conventional, its crises arranged, its conflicts performed, and its metaphysics ideological.

45

To regard society as a species of culture is not to overthrow or even alter society, but only to eliminate its perceived necessity. Infinite players have rules; they just do not forget that rules are an expression of agreement and not a requirement for agreement.

Culture is not therefore mere disorder. Infinite players never understand their culture as the composite of all that they choose individually to do, but as the congruence of all that they choose to do with each other. Because there is no congruence without the decision to have one, all cultural congruence is under constant revision. No sooner did the Renaissance begin than it began to change. Indeed, the Renaissance was not something apart from its change; it was itself a certain persistent and congruent evolution.

For this reason it can be said that where a society is defined by its *boundaries,* a culture is defined by its *horizon.*

A boundary is a phenomenon of opposition. It is the meeting place of hostile forces. Where nothing opposes there can be no boundary. One cannot move beyond a boundary without being resisted.

This is why patriotism—that is, the desire to protect the power in a society by way of increasing the power of a society— is inherently belligerent. Since there can be no prizes without a society, no society without opponents, patriots must create enemies before we can require protection from them. Patriots can flourish only where boundaries are well-defined, hostile, and dangerous. The spirit of patriotism is therefore characteristically associated with the military or other modes of international conflict.

Because patriotism is the desire to contain all other finite games within itself—that is, to embrace all horizons within a single boundary—it is inherently evil.

A horizon is a phenomenon of vision. One cannot look at the horizon; it is simply the point beyond which we cannot see. There is nothing in the horizon itself, however, that limits vision, for the horizon opens onto all that lies beyond itself. What limits vision is rather the incompleteness of that vision.

One never reaches a horizon. It is not a line; it has no place; it encloses no field; its location is always relative to the view. To move toward a horizon is simply to have a new horizon. One can therefore never be close to one's horizon, though one may certainly have a short range of vision, a narrow horizon.

We are never somewhere in relation to the horizon since the horizon moves with our vision. We can only be somewhere by turning away from the horizon, by replacing vision with opposition, by declaring the place on which we stand to be timeless—a sacred region, a holy land, a body of truth, a code of inviolable commandments. To be somewhere is to absolutize time, space, and number.

Every move an infinite player makes is toward the horizon. Every move made by a finite player is within a boundary. Every moment of an infinite game therefore presents a new vision, a new range of possibilities. The Renaissance, like all genuine cultural phenomena, was not an effort to promote one or another vision. It was an effort to find visions that promised still more vision.

Who lives horizonally is never somewhere, but always in passage.

46

Since culture is horizonal it is not restricted by time or space.

To the degree that the Renaissance was true culture it has not ended. Anyone may enter into its mode of renewing vision. This does not mean that we repeat what was done. To enter a culture is not to do what the others do, but to do whatever one does with the others.

This is why every new participant in a culture both enters into an existing context and simultaneously changes that context. Each new speaker of its language both learns the language and alters it. Each new adoption of a tradition makes it a new tradition—just as the family into which a child is born existed prior to that birth, but is nonetheless a new family after the birth.

The reciprocity of this transformation has no respect to time. The fact that the Renaissance began in the fourteenth or fifteenth century has nothing to do with its capacity for changing our horizon. This reciprocity works backward as well as forward. Each person whose horizon is affected by the Renaissance affects the horizon of the Renaissance in turn. Any culture that continues to influence our vision continues to grow in the very exercise of that influence.

47

Since a culture is not anything persons do, but anything they do with each other, we may say that a culture comes into being whenever persons choose to be a people. It is as a people that they arrange their rules with each other, their moralities, their modes of communication.

Properly speaking, the Renaissance is not a period but a people, moreover, a people without a boundary, and therefore without an enemy. The Renaissance is not against anyone. Whoever is not of the Renaissance cannot go out to oppose it, for they will find only an invitation to join the people it is.

A culture is sometimes opposed by suppressing its ideas, its works, even its language. This is a common strategy of a society afraid of the culture growing within its boundaries. But it is a strategy certain to fail, because it confuses the creative activity (poiesis) with the product (poiema) of that activity.

Societies characteristically separate the ideas from their thinkers, the poiema from its poietes. A society abstracts its thought and grants power to certain ideas as though they had an existence of their own independent from those who think them, even those who first produced them. In fact, a society is likely to have an idea of itself that no thinker may challenge or revise. Abstracted thought—thought without a thinker—is metaphysics. A society's metaphysics is its ideology: theories that present themselves as the product of these people or those. The Renaissance had no ideology.

Inasmuch as it has no metaphysics, a people is not threatened when its apparent society is threatened, or altered, or even destroyed. The manipulation of the government, the laws, the enforcement functions of a state either by persons within the society (through usurpation or abuse of power) or by persons without (in other states) cannot in itself affect the decision of a people to be a people.

A people, as a people, has nothing to defend. In the same way a people has nothing and no one to attack. One cannot be free by opposing another. My freedom does not depend on your loss of freedom. On the contrary, since freedom is never freedom from society, but freedom for it, my freedom inherently affirms yours.

A people has no enemies.

48

For a bounded, metaphysically veiled, and destined society, enemies are necessary, conflict inevitable, and war likely.

War is not an act of unchecked ruthlessness but a declared contest between bounded societies, or states. If a state has no enemies it has no boundaries. To keep its definitions clear a state must stimulate danger to itself. Under the constant danger of war the people of a state are far more attentive and obedient to the finite structures of their society: "just as the blowing of the winds preserves the sea from the foulness which would be the result of a prolonged calm, so also corruption in nations would be the product of prolonged, let alone 'perpetual' peace" (Hegel).

War presents itself as necessary for self-protection, when in fact it is necessary for self-identification.

If it is the impulse of a finite player to go against another nation in war, it is the design of an infinite player to oppose war within a nation.

If as a people infinite players cannot go to war against a people, they can act against war itself within whatever state they happen to reside. In one way their opposition to war resembles that of finite players: Each is opposed to the existence of a state. But their reasons and the strategies for attempting to eliminate states are radically different. Finite players go to war against states because they *endanger* boundaries; infinite players oppose states because they *engender* boundaries.

The strategy of finite players is to kill a state by killing the people who invented it. Infinite players, however, understanding war to be a conflict between states, conclude that states can have only states as enemies; they cannot have persons as enemies. "Sometimes it is possible to kill a state without killing a single one of its members; and war gives no right which is not necessary to the gaining of its object" (Rousseau). For infinite players, if it is possible to wage a war without killing a single person, then it is possible to wage war *only* without killing a single person.

For infinite players the chief difficulty with finite players' commitment to war is not, however, that persons are killed.

Indeed, finite players themselves often genuinely regret this and do as little killing as possible. The difficulty is that such warfare has in it the contradiction of all finite play. Winning a war can be as destructive as losing one, for if boundaries lose their clarity, as they do in a decisive victory, the state loses its identity. Just as Alexander wept upon learning he had no more enemies to conquer, finite players come to rue their victories unless they see them quickly challenged by new danger. A war fought to end all wars, in the strategy of finite play, only breeds universal warfare.

The strategy of infinite players is horizonal. They do not go to meet putative enemies with power and violence, but with poiesis and vision. They invite them to become a people in passage. Infinite players do not rise to meet arms with arms; instead, they make use of laughter, vision, and surprise to engage the state and put its boundaries back into play.

What will undo any boundary is the awareness that it is our vision, and not what we are viewing, that is limited.

49

Plato suggested that some of the poets be driven out of the Republic because they had the power to weaken the guardians. Poets can make it impossible to have a war—unless they tell stories that agree with the "general line" established by the state. Poets who have no metaphysics, and therefore no political line, make war impossible because they have the irresistible ability to show the guardians that what seems necessary is only possible.

The danger of the poets, for Plato, is that they can imitate so well that it is difficult to see what is true and what is merely invented. Since reality cannot be invented, but only discovered through the exercise of reason—according to Plato—all poets must be put into the service of reason. The poets are to surround the citizens of the Republic with such art as will "lead them unawares from childhood to love of, resemblance to, and harmony with, the beauty of reason."

The use of the word "unawares" shows Plato's intention to keep the metaphysical veil intact. Those who are being led to reason cannot be aware of it. They must be led to it without choosing it. Plato asks his poets not to create, but to deceive.

True poets lead no one unawares. It is nothing other than awareness that poets—that is, creators of all sorts—seek. They do not display their art so as to make it appear real; they display the real in a way that reveals it to be art.

We must remind ourselves, to be sure, that Plato was himself an artist, a poietes. His Republic was an invention. So were the theory of forms and the idea of the Good. Since all veiling is self-veiling, we cannot help but think that behind the rational metaphysician, philosophy's great Master Player, stood Plato the poet, fully aware that the entire opus was an act of play, an invitation to readers not to reproduce the truth but to take his inventions into their own play, establishing the continuity of his art by changing it.

50

We can find metaphysicians thinking, but we cannot find metaphysicians in their thinking. When we separate the metaphysics from the thinker we have an abstraction, the deathless shadow of a once living act. It is no longer what someone is saying but what someone has said. When metaphysics is most successful on its own terms, it leaves its listeners in silence, certainly not in laughter.

Metaphysics is *about* the real but is abstract. Poetry is the *making* (poiesis) of the real and is concrete. Whenever *what is made* (poiema) is separated from the *maker* (poietes), it becomes metaphysical. As it stands there, and as the voice of the poietes is no longer listened to, the poiema is an object to be studied, not an act to be learned. One cannot learn an object, but only the poiesis, or the act of creating objects. To separate the poiema from poiesis, the created object from the creative act, is the essence of the theatrical.

Poets cannot kill; they die. Metaphysics cannot die; it kills.

THREE

I Am the Genius of Myself

51

I AM THE GENIUS of myself, the poietes who composes the sentences I speak and the actions I take. It is I, not the mind, that thinks. It is I, not the will, that acts. It is I, not the nervous system, that feels.

When I speak as the genius I am, I speak these words for the first time. To repeat words is to speak them as though another were saying them, in which case I am not saying them. To be the genius of my speech is to be the origin of my words, to say them for the first, and last, time. Even to repeat my own words is to say them as though I were another person in another time and place.

When I forsake my genius and speak to you as though I were another, I also speak to you as someone you are not and somewhere you are not. I address you as audience, and do not expect you to respond as the genius you are.

Hamlet was not reading when he said he was reading words; neither do we act when we perform actions, nor think when we entertain thoughts. A dog taught the action of shaking hands does not shake your hand. A robot can say words but cannot say them to you.

Since being your own genius is dramatic, it has all the paradox of infinite play: You can have what you have only by releasing it to others. The sounds of the words you speak may lie on your own lips, but if you do not relinquish them entirely to a listener they never become words, and you say nothing at all. The words die with the sound. Spoken to me, your words become mine to do with as I please. As the genius of your words, you lose all authority over them. So too with thoughts. However you consider them your own, you cannot think the thoughts themselves, but only what they are *about*. You cannot think thoughts any more than you can act actions. If you do not truly speak the words that reside entirely in their own sound, neither can you think that which remains thought or can be translated back into thought. In thinking you cast thoughts beyond themselves, surrendering them to that which they cannot be.

The paradox of genius exposes us directly to the dynamic of open reciprocity, for if you are the genius of what you say to me, I am the genius of what I hear you say. What you say originally I can hear only originally. As you surrender the sound on your lips, I surrender the sound in my ear. Each of us has relinquished to the other what has been relinquished to the other.

This does not mean that speech has come to nothing. On the contrary, it has become speech that invites speech. When the genius of speech is abandoned, words are said not originally but repetitively. To repeat words, even our own, is to contain them in their own sound. Veiled speech is that spoken as though we have forgotten we are its originators.

To speak, or act, or think originally is to erase the boundary of the self. It is to leave behind the territorial personality. A genius does not have a mind full of thoughts but is the thinker of thoughts, and is the center of a field of vision. It is a field of vision, however, that is recognized as a field of vision only when we see that it includes within itself the original centers of other fields of vision.

This does not mean that I can see what you see. On the contrary, *it is because I cannot see what you see that I can see at all.* The discovery that you are the unrepeatable center of your own vision is simultaneous with the discovery that I am the center of my own.

52

As the geniuses we are, we do not *look* but *see*.

To look at something is to look at it within its limitations. I look at what is marked off, at what stands apart from other things. But things do not have their own limitations. Nothing limits itself. The sea gulls circling on the invisible currents, the cat on my desk, the siren of a distant ambulance are not somehow distinct from the environment; they are the environment. To look at them I must look for what I take them to be. I was not looking at the sea gulls as though it was the sea gulls who happened to be there—I was looking for something to make this example. I might have seen them as a sign that land is not far, or that the sea is not far; I could have been looking for a form to reproduce on a canvas or in a poem. To look at is to look for. It is to bring the limitations with us. "Nature has no outline. Imagination has" (Blake).

If to look is to look at what is contained within its limitations, to see is to see the limitations themselves. Each new school of painting is new not because it now contains subject matter ignored in earlier work, but because it sees the limitations previous artists imposed on their subject matter but could not see themselves. The earlier artists worked within the outlines they imagined; the later reworked their imaginations.

To look is a territorial activity. It is to observe one thing after another within a bounded space—as though in time it can all be seen. Academic fields are such territories. Sometimes everything in a field finally does get looked at and defined— that is, placed in its proper location. Mechanics and rhetoric are such fields. Physics may prove to be. Biological mysteries fall away at an astonishing rate. It becomes increasingly difficult to find something new to look at.

When we pass from looking to seeing, we do not therefore lose our sight of the objects observed. Seeing, in fact, does not disturb our looking at all. It rather places us in that territory as its genius, aware that our imagination does not create within its outlines but creates the outlines themselves. The physicist who sees speaks physics with us, inviting us to see that the things we thought were there are not things at all. By learning new limitations from such a person, we learn not only what to look for with them but also how to see the way we use limitations. A physics so taught becomes poiesis.

53

To be the genius of myself is not to bring myself into being. As the origin of myself I am not also the cause of myself, as though I were the product of my own action. But then neither am I the product of any other action. My parents may have wanted a child, but they could not have wanted me.

I am both the outcome of my past and the transformation of my past. To be related to the past as its outcome is to stand in causal continuity with it. Such a relation can be accounted for in scientific explanation. I can be said to be the result of precise genetic influence. The date and place of my birth are matters of causal necessity; I had no part in deciding either. Neither could anyone else have chosen them. My birth, when understood in terms of causal continuity, marks no absolute beginning. It marks nothing at all except an arbitrary point in an unbroken process. Causally speaking, there is nothing new here, only the kinds of change that conform to the known laws of nature.

Speaking in purely causal terms, I cannot say I was born; I should say rather that I have emerged as a phase in the process of reproduction. A reproduction is a repetition, a recurrence of that which has been. Birth, on the other hand, in causal terms, is all discontinuity. It has its beginning in itself, and can be caused by nothing. It makes no sense to say, "I was reproduced on this date and in this place." To say "I was born" is to speak of myself as having an uncaused point of departure within the realm of the continuous, an absolute beginning not comprehensible to the explanatory intelligence.

As such a phenomenon birth repeats nothing; it is not the outcome of the past but the recasting of a drama already under way. A birth is an event in the ongoing history of a family, even the history of a culture. The radical originality of a birth announces itself in the way it brings the dramatic into conflict with the theatrical in cultural or family history.

Theatrically, my birth is an event of plotted repetition. I am born as another member of my family and my culture. Who I am is a question already answered by the content and character of a tradition. Dramatically, my birth is the rupture of that repetitive sequence, an event certain to change what the past has meant. In this case the character of a tradition is determined by who I am. Dramatically speaking, every birth is the birth of genius.

The drama under way at the time of my birth is moved forward to new possibilities by the appearance of a new genius within it. It is a drama, however, already peopled by finite players attempting to forget, playing for keeps. If I am born into, and add to, the culture of a family, I am also a product and a citizen of its politics. I first experience the conflict between the theatrical and the dramatic in the felt pressure to take up one of the roles prepared for me: eldest son, favorite daughter, heir to the family's honor, avenger of its losses.

Each of these roles comes, of course, with a script, one whose lines a person might easily spend a lifetime repeating, while intentionally forgetting, or repressing, the fact that it is but a learned script. Such a person "is obliged to *repeat* the repressed material as a contemporary experience instead of, as the physician would prefer to see, *remembering* it as something belonging to the past" (Freud). It is the genius in us who knows that the past is most definitely past, and therefore not forever sealed but forever open to creative reinterpretation.

54

Not allowing the past to be past may be the primary source for the seriousness of finite players. Inasmuch as finite play always has its audience, it is the audience to whom the finite player intends to be known as winner. The finite player, in other words, must not only have an audience but must have an audience to convince.

Just as the titles of winners are worthless unless they are visible to others, there is a kind of antititle that attaches to invisibility. To the degree that we are invisible we have a past that has condemned us to oblivion. It is as though we have somehow been overlooked, even forgotten, by our chosen audience. If it is the winners who are presently visible, it is the losers who are invisibly past.

As we enter into finite play—not playfully, but seriously—we come before an audience conscious that we bear the antititles of invisibility. We feel the need, therefore, to prove to them that we are not what we think they think we are or, more precisely, that we were not who we think the audience thinks we were.

As with all finite play, an acute contradiction quickly develops at the heart of this attempt. As finite players we will not enter the game with sufficient desire to win unless we are ourselves convinced by the very audience we intend to convince. That is, *unless we believe we actually are the losers the audience sees us to be, we will not have the necessary desire to win.* The more negatively we assess ourselves, the more we strive to reverse the negative judgment of others. The outcome brings the contradiction to perfection: by proving to the audience they were wrong, we prove to ourselves the audience was right.

The more we are recognized as winners, the more we know ourselves to be losers. That is why it is rare for the winners of highly coveted and publicized prizes to settle for their titles and retire. Winners, especially celebrated winners, must prove repeatedly they are winners. The script must be played over and over again. Titles must be defended by new contests. No one is ever wealthy enough, honored enough, applauded enough. On the contrary, the visibility of our victories only tightens the grip of the failures in our invisible past.

So crucial is this power of the past to finite play that we must find ways of remembering that we have been forgotten to sustain our interest in the struggle. There is a humiliating memory at the bottom of all serious conflicts. "Remember the Alamo!" "Remember the Maine!" "Remember Pearl Harbor!" These are the cries that carried Americans into several wars. Having once been insulted by Athens, the great Persian Emperor Darius renewed his appetite for war by having a page follow him about to whisper in his ear, "Sire, remember the Athenians."

Indeed, it is only by remembering what we have forgotten that we can enter into competition with sufficient intensity to be able to forget we have forgotten the character of all play: Whoever *must* play cannot *play*.

Whenever we act as the genius of ourselves, it will be in the spirit of allowing the past to be past. It is the genius in us who is capable of ridding us of resentment by exercising what Nietzsche called the "faculty of oblivion," not as a way of denying the past but as a way of reshaping it through our own originality. Then we forget that we have been forgotten by an audience, and remember that we have forgotten our freedom to play.

55

If in the culture into which we are born there are always persons who will urge us to theatricalize our lives by supplying us with a repeatable past, there will also be persons (possibly the same ones) in whose presence we learn to prepare ourselves for surprise. It is in the presence of such persons that we first recognize ourselves as the geniuses we are.

These persons do not give us our genius or produce it in us. In no way is the source of genius external to itself; never is a child moved to genius. Genius arises with *touch*. Touch is a characteristically paradoxical phenomenon of infinite play.

I am not touched by an other when the distance between us is reduced to zero. I am touched only if I respond from my own center—that is, spontaneously, originally. But you do not touch me except from your own center, out of your own genius. Touching is always reciprocal. You cannot touch me unless I touch you in response.

The opposite of touching is *moving*. You move me by pressing me from without toward a place you have already foreseen and perhaps prepared. It is a staged action that succeeds only if in moving me you remain unmoved yourself. I can be moved to tears by skilled performances and heart-rending newspaper accounts, or moved to passion by political manifestos and narratives of heroic achievement—but in each case I am moved according to a formula or design to which the actor or agent is immune. When actors bring themselves to tears *by* their performance, and not *as* their performance, they have failed their craft; they have become theatrically inept.

This means that we can be moved only by persons who are not what they are; we can be moved only when we are not who we are, but are what we cannot be.

When I am touched, I am touched only as the person I am behind all the theatrical masks, but at the same time I am changed from within—and whoever touches me is touched as well. We do not touch by design. Indeed, all designs are shattered by touching. Whoever touches and whoever is touched cannot but be surprised. (The unpredictability of this phenomenon is reflected in our reference to the insane as "touched.")

We can be moved only by way of our veils. We are touched through our veils.

The character of touching can be seen quite clearly in the way infinite players understand both healing and sexuality.

If to be touched is to respond from one's center, it is also to respond as a whole person. To be whole is to be hale, or healthy. In sum, whoever is touched is healed.

The finite player's interest is not in being healed, or made whole, but in being cured, or made functional. Healing restores me to play, curing restores me to competition in one or another game.

Physicians who cure must abstract persons into functions. They treat the illness, not the person. And persons willfully present themselves as functions. Indeed, what sustains the enormous size and cost of the curing professions is the widespread desire to see oneself as a function, or a collection of functions. To be ill is to be dysfunctional; to be dysfunctional is to be unable to compete in one's preferred contests. It is a kind of death, an inability to acquire titles. The ill become invisible. Illness always has the smell of death about it: Either it may lead to death, or it leads to the death of a person as competitor. The dread of illness is the dread of losing.

One is never ill in general. One is always ill with relation to some bounded activity. It is not cancer that makes me ill. It is because I cannot work, or run, or swallow that I am ill with cancer. The loss of function, the obstruction of an activity, cannot in itself destroy my health. I am too heavy to fly by flapping my arms, but I do not for that reason complain of being sick with weight. However, if I desired to be a fashion model, a dancer, or a jockey, I would consider excessive weight to be a kind of disease and would be likely to consult a doctor, a nutritionist, or another specialist to be cured of it.

When I am healed I am restored to my center in a way that my freedom as a person is not compromised by my loss of functions. This means that the illness need not be eliminated before I can be healed. I am not free to the degree that I can overcome my infirmities, but only to the degree that I can put my infirmities into play. I am cured of my illness; I am healed with my illness.

Healing, of course, has all the reciprocity of touching. Just as I cannot touch myself, I cannot heal myself. But healing requires no specialists, only those who can come to us out of their own center, and who are prepared to be healed themselves.

57

Sexuality for the infinite player is entirely a matter of touch. One cannot touch without touching sexually.

Because sexuality is a drama of origins, it gives full expression to the genius you are and to the genius of others who participate in that drama. This throws a high challenge before the political ideologue. Aware that genuine sexual expression is at least as dangerous to society as genuine artistic expression, the sexual metaphysician can appeal to at least two powerful solutions. One is to treat sexuality as a process of reproduction; another is to place it in the area of feeling and behavior.

Although reproduction is a process that operates by way of our bodies, it nonetheless operates autonomously. Like every other natural process it is a phenomenon of causal continuity, having no inherent beginning or end. Therefore we cannot be said to *initiate* the process by any act of our own. We can only be carried along by it, inasmuch as conception occurs only when all the necessary conditions have been met by the parenting couple. No one conceives a child; a child is conceived in the conjunction of sperm and ovum. The mother does not give birth to a child; the mother is where the birth occurs.

The metaphysics of sexuality, applying to this solution, can therefore draw a boundary line around sexual activity that leaves the genius of parenting altogether outside it. Thus the familiar view of some Christian theologians who say that the only end of the sexual act is procreation. But this metaphysics, committed as it is to the continuity of the process, also leaves the genius of the child entirely outside it. Thus the familiar view of theologians who say that the end of childbirth is to provide citizens for the kingdom of God. Metaphysically understood, sexuality has nothing to do with our existence as persons, for it views persons as expressions of sexuality, and not sexuality as the expression of persons.

The second way of veiling genuine sexuality is to regard it as a feeling or as a kind of behavior. In either case it has the character of something under observation. Even if it is our own sexuality we are concerned with, we can still look on it from without, making an assessment of it as though it were of another person. We ask ourselves and each other whether certain behavior is acceptable or desirable; we are puzzled over the proper response to sexual feelings—ours or another's. Sexuality can in this way be dealt with as a societal phenomenon, regulated and managed according to the prevailing ideology. Sexual rebels, violators of the sexual taboos, do not weaken this ideology but affirm it as the rules of finite play.

It is convenient to think that sexual misfits violate rules. The matter is subtler by far. They are not concerned to oppose the rules themselves but to engage in competitive struggle by way of those rules. Sexual attractiveness, or sexiness, is effective only to the degree that someone is offended by it.

Pornography is exciting only so far as it reveals something forbidden, something otherwise unseeable. Thus the mandatory hostility in it, the quality of shock and violence.

Because sexuality is so rich in the mystery of origin, it becomes a region of human action deeply shaped by resentment, where participants play out a manifold strategy of hostile encounters. The players in finite sexuality not only require the offended resistance of those who refuse to join them in their play, they require the resistance of those who do join them.

Sexual plotting on the part of one player is in fact stimulated by disinterest or fear or loathing on the part of the other. A Master Player of finite sexuality chooses not to take these attitudes as a way of refusing the sexual game, but takes them to be part of the game. Thus my indifference or revulsion to your sexuality becomes in your masterful play a *sexual* indifference, a *sexual* revulsion. Suddenly I am no longer indifferent to your game, but indifferent to you within your game, and have ipso facto made myself your opponent. This is the plot of the classical pulp novel and of Hollywood romance: indifferent girl won by ardent boy.

The profound seriousness of such sexual play is seen in the unique nature of the prize that goes to the winner. What one wants in the sexual contest is not just to have defeated the other, but to have the defeated other. *Sexuality is the only finite game in which the winner's prize is the defeated opponent.*

Sexual titles, like all other titles, have appropriately conspicuous emblems. However, only in sexuality do persons themselves become property. In slavery or wage labor what we possess is not the persons of the slaves or the workers, but the products of their labors. In this case, to use Marx's phrase, persons are abstracted from their labor. But in sexuality persons are abstracted from themselves. The seduced opponent is so displayed as to draw public attention to the seducer's triumph. In the complex plotting of sexual encounter it is by no means uncommon for the partners to have played a double game in which each is winner and loser, and each is an emblem for the other's seductive power.

A society shows its mastery in the management of sexuality not when it sets out unambiguous standards for sexual behavior or prescribed attitudes toward sexual feelings, but when it institutionalizes the emblematic display of sexual conquest. These institutions can be as varied as burning widows alive on the funeral pyres of their husbands or requiring the high visibility of a spouse at an elected official's inauguration.

Finite sexuality is a form of theater in which the distance between persons is regularly reduced to zero but in which neither touches the other.

58

Insofar as sexuality is a drama of origin it is original to society and not derivative of it. It is therefore somewhat misleading to describe society as a regulator of finite sexual play. It is more the case that finite sexuality shapes society than is shaped by it. Only to a limited extent do we take on the sexual roles assigned us by society. Much more frequently we enter into societal arrangements by way of sexual roles. (For example, we are more likely to refer to the king as the father of the country than we are to refer to the father as king of the family.) While society does serve a regulatory function, it is probably more correctly understood as sexuality making use of society to regulate itself.

This means that society plays little or no role in either

causing or preventing sexual tensions. On the contrary, society absorbs sexual tensions into all of its structures. It becomes the larger theater for playing out the patterns of resentment learned in the family. Society is where we prove to parents qua audience that we are not what we thought they thought we were. Since the emphasis in this relationship is not on what our parents thought of us but on what we thought they thought, they become an audience that easily survives their physical absence or death. Moreover, for the same reason they become an audience whose definitive approval we can never win.

To use Freud's famous phrase, the civilized are, therefore, the discontent. We do not become losers in civilization but become civilized as losers. The collective result of this ineradicable sense of failure is that civilizations take on the spirit of resentment. Acutely sensitive to an imagined audience, they are easily offended by other civilizations. Indeed, even the most powerful societies can be embarrassed by the weakest: the Soviet Union by Afghanistan, Great Britain by Argentina, the United States by Grenada.

This is also why the only true revolutionary act is not the overthrow of the father by the son—which only reinforces the existing patterns of resentment—but the restoration of genius to sexuality. It is by no means an accident that the only successful attempt of the American citizenry to force the ending of a foreign war occurred simultaneously with a wide revision in sexual attitudes. The civilization quickly recovered from this threat, however, by tempting these revolutionaries into a new sexual politics, one of societal standoff, where sexual genius is confused with such struggles as the passage of the Equal Rights Amendment and the election of women to national office.

There is one other way in which society is shaped by the

tensions of finite sexuality: in its orientation toward property. Since sexuality is the only finite game in which the winner's prize is the loser, the most desirable form of property is the publicly acknowledged possession of another's person, a relationship to which the possessed must of course freely consent. All other forms of property are considerably less desirable, even when they are vast in quantity. The true value of my property, in fact, varies not with its monetary worth but with its effectiveness in winning for me the declaration that I am the Master Player in our game with each other.

The most serious struggles are those for sexual property. For this wars are fought, lives are generously risked, great schemes are initiated. However, who wins empire, fortune, and fame but loses in love has lost in everything.

59

Because finite, or veiled, sexuality is one or another struggle which its participants mean to win, it is oriented toward moments, outcomes, final scenes. Like all finite play it proceeds largely by deception. Sexual desires are usually not directly announced but concealed under a series of feints, gestures, styles of dress, and showy behavior. Seductions are staged, scripted, costumed. Certain responses are sought, plots are developed. In skillful seductions delays are employed, special circumstances and settings are arranged.

Seductions are designed to come to an end. Time runs out. The play is finished. All that remains is recollection, the memory of a moment, and perhaps a longing for its repetition. Seductions cannot be repeated. Once one has won or lost in a particular finite game, the game cannot be played over.

Moments once reached cannot be reached again. Lovers often sustain vivid reminders of extraordinary moments, but they are reminded at the same time of their impotence in recreating them. The appetite for novelty in lovemaking—new positions, the use of drugs, exotic surroundings, additional partners— is only a search for new moments that can live on only in recollection.

As with all finite play, the goal of veiled sexuality is to bring itself to an end.

60

By contrast, infinite players have no interest in seduction or in restricting the freedom of another to one's own bound- aries of play. Infinite players recognize choice in all aspects of sexuality. They may see in themselves and in others, for example, the infant's desire to compete for the mother, but they also see that there is neither physiological nor societal destiny in sexual patterns. Who chooses to compete with an- other can also choose to play with another.

Sexuality is not a bounded phenomenon but a horizonal phenomenon for infinite players. One can never say, therefore, that an infinite player is homosexual, or heterosexual, or celi- bate, or adulterous, or faithful—because each of these defini- tions has to do with boundaries, with circumscribed areas and styles of play. Infinite players do not play within sexual boundaries, but with sexual boundaries. They are concerned not with power but with vision.

In their sexual play they suffer others, allow them to be as they are. Suffering others, they open themselves. Open, they learn both about others and about themselves. Learning, they grow. What they learn is not about sexuality, but how to be more concretely and originally themselves, to be the genius of their own actions, to be whole.

Moving therefore from an original center, the sexual engagements of infinite players have no standards, no ideals, no marks of success or failure. Neither orgasm nor conception is a goal in their play, although either may be part of the play.

61

There is nothing hidden in infinite sexuality. Sexual desire is exposed *as* sexual desire and is never therefore serious. Its satisfaction is never an achievement, but an act in a continuing relationship, and therefore joyous. Its lack of satisfaction is never a failure, but only a matter to be taken on into further play.

Infinite lovers may or may not have a family. Rousseau said the only human institution that is not conventional is the family, which for a brief time is required by nature. Rousseau erred. No family is united by natural or any other kind of necessity. Families can convene only out of choice. The family of infinite lovers has this difference, that it is self-evidently chosen. It is a progressive work of unveiling. Fathering and mothering are roles freely assumed but always with the design of showing them to be theatrical. It is the intention of parents in such families to make it plain to their children that they all play cultural and not societal roles, that they are only roles, and that they are all truly concrete persons behind them. Therefore, children also learn that they have a family only by choosing to have it, by a collective act to be a family with each other.

Infinite sexuality does not focus its attention on certain parts or regions of the body. Infinite lovers have no "private parts." They do not regard their bodies as having secret zones that can be exposed or made accessible to others for special favors. It is not their bodies but their persons they make accessible to others.

The paradox of infinite sexuality is that by regarding sexuality as an expression of the person and not the body, it becomes fully embodied play. It becomes a drama of touching.

The triumph of finite sexuality is to be liberated from play into the body. The essence of infinite sexuality is to be liberated into play with the body. In finite sexuality I expect to relate to you as a body; in infinite sexuality I expect to relate to you in your body.

Infinite lovers conform to the sexual expectations of others in a way that does not expose something hidden, but unveils something in plain sight: that sexual engagement is a poiesis of free persons. In this exposure they emerge as the persons they are. They meet others with their limitations, and not within their limitations. In doing so they expect to be transformed—and are transformed.

FOUR

A Finite Game Occurs Within a World

63

A FINITE GAME occurs within a world. The fact that it must be limited temporally, numerically, and spatially means that there is something against which the limits stand. There is an *outside* to every finite game. Its limits are meaningless unless there is something to be limited, unless there is a larger space, a longer time, a greater number of possible competitors.

There is nothing about a finite game, in itself, that determines *at what time* it is to be played, or *by whom*, or *where*.

The rules of a finite game will indicate the temporal, spatial, and numerical nature of the game itself; that, for example, it will last sixty minutes, will be played on a field 100 yards in length, and by two teams of eleven players each. But the rules do not, and cannot, determine the date, the location, and the specific participants. There is nothing in the rules that requires professional teams composed of certain persons, earning salaries of specified amounts, joining at the end of each season in a national championship. The rules for the practice of medicine or for the exercise of the office of the Bishop of Rome do not indicate which persons are to enter these games; which *kinds* of persons, yes, but *never the names of anyone*.

A world provides an absolute reference without which the time, place, and participants make no sense.

Whatever occurs within a game is *relatively* intelligible with reference to whatever else has happened inside its boundaries, but it is *absolutely* intelligible with reference to that world for the sake of which its boundaries exist.

It is relatively intelligible to have won an election for the presidency by a few thousand votes after a campaign of about ten months; it is absolutely intelligible to be the sixteenth President of the United States, a society unambiguously marked off from all the rest of the world by its declared boundaries, in the 1,860th year of that world's history.

We cannot have a precise understanding of what it means to be the winner of a contest until we can place the game in the absolute dimensions of a world.

64

World exists in the form of audience. A world is not all that is the case, but that which determines all that is the case.

An audience consists of persons observing a contest without participating in it.

No one determines who an audience will be. No exercise of power can make a world. A world must be its own spontaneous source. "A world worlds" (Heidegger). Who *must* be a world cannot *be* a world.

The number of persons who join an audience is irrelevant. So is the time and space in which an audience occurs. The temporal and spatial boundaries of a finite game must be absolute—in relation to an audience or a world. But when and where a world occurs, and whom it includes, is of no importance. One does not say, "I was in the world, or audience, on November 22, 1963," but rather, "I was just getting out of the car thinking about what to cook for dinner when I heard that the President had been shot." An audience does not receive its identity according to the persons within it, but according to the events it observes. Those who remember that day remember precisely what they were doing in the early afternoon of that day, not because it was the 22d of November, but because it was at that moment that they became audience to the events of that day.

If the boundaries of an audience are irrelevant, what is relevant is the unity of the audience. They must be a singular entity, bound in their desire to see who will win the contest before them. Anyone for whom this desire is not primary is not in the audience for that contest, and is not a person in that world.

The fact that a finite game needs an audience before which it can be played, and the fact that an audience needs to be singularly absorbed in the events before it, show the crucial reciprocity of finite play and the world. Finite players need the world to provide an absolute reference for understanding themselves; simultaneously, the world needs the theater of finite play to remain a world. George Eliot's villainous character, Grandcourt, "did not care a languid curse for anyone's admiration; but this state of non-caring, just as much as desire, required its related object—namely, a world of admiring and envying spectators: for if you are fond of looking stonily at smiling persons, the persons must be there and they must smile."

We are players in search of a world as often as we are world in search of players, and sometimes we are both at once. Some worlds pass quickly into existence, and quickly out of it. Some sustain themselves for longer periods, but no world lasts forever.

65

There is an indefinite number of worlds.

66

The reciprocity of game and world has another, deeper effect on the persons involved. Because the seriousness of finite play derives from the players' need to correct another's putative assessment of themselves, there is no requirement that the audience be physically present, since players are already their own audience. Just as in finite sexuality where the absence or death of parents has no effect on the child's determination to prove them wrong, finite players become their own hostile observers in the very act of competing.

I cannot be a finite player without being divided against myself.

A similar dynamic is found in the audience. When sufficiently oblivious to their status as audience, the observers of a finite game become so absorbed in its conduct that they lose the sense of distance between themselves and the players. It is they, quite as much as the players, who win or lose. For this reason the audience absorbs in itself the same politics of resentment that moves players to show they are not what they think others think they are. The audience is under the same constraint to disprove this judgment.

When we ask where an audience will find its own audience, we discover the division inherent in all audiences. Each side of a conflict comes with its own partisan observers. Inasmuch as the conflict is expressed within the bounded playing of a game, the audience is unified—but its unity consists in its opposition to itself.

We cannot become a world without being divided against ourselves.

67

Occurring before a world, theatrically, a finite game occurs within time. Because it has its boundaries, its beginning and end, within the absolute temporal limits established by a world, time for a finite player runs out; it is used up. It is a diminishing quantity.

A finite game does not have its own time. It exists in a world's time. An audience allows players only so much time to win their titles.

Early in a game time seems abundant, and there appears a greater freedom to develop future strategies. Late in a game, time is rapidly being consumed. As choices become more limited they become more important. Errors are more disastrous.

We look on childhood and youth as those "times of life" rich with possibility only because there still seem to remain so many paths open to a successful outcome. Each year that passes, however, increases the competitive value of making strategically correct decisions. The errors of childhood can be more easily amended than those of adulthood.

For the finite player in us freedom is a function of time. We must have time to be free.

The passage of time is always relative to that which does not pass, to the timeless. Victories occur in time, but the titles won in them are timeless. Titles neither age nor die.

The points of reference for all finite history are signal triumphs meant never to be forgotten: establishment of the throne of David, the birth of the Savior, the journey to Medina, the battle of Hastings, the American, French, Russian, Chinese, and Cuban revolutions.

Time divided into periods is theatrical time. The lapse of time between the opening and closing of an era is a scene between curtains. It is not a time lived, but a time viewed— by both players and audience. The periodization of time presupposes a viewer existing outside the boundaries of play, able to see the beginning and the end simultaneously.

The outcome of a finite game is the past waiting to happen. Whoever plays toward a certain outcome desires a particular past. By competing for a future prize, finite players compete for a prized past.

68

The infinite player in us does not consume time but generates it. Because infinite play is dramatic and has no scripted conclusion, its time is time lived and not time viewed.

As an infinite player one is neither young nor old, for one does not live in the time of another. There is therefore no external measure of an infinite player's temporality. Time does not pass for an infinite player. Each moment of time is a beginning.

Each moment is not the beginning of a *period of time*. It is the beginning of an event that gives the time within it its specific quality. For an infinite player there is no such thing as an hour of time. There can be an hour of love, or a day of grieving, or a season of learning, or a period of labor.

An infinite player does not begin working for the purpose

of filling up a period of time with work, but for the purpose of filling work with time. Work is not an infinite player's way of passing time, but of engendering possibility. Work is not a way of arriving at a desired present and securing it against an unpredictable future, but of moving toward a future which itself has a future.

Infinite players cannot say how much they have completed in their work or love or quarreling, but only that much remains incomplete in it. They are not concerned to determine when it is over, but only what comes of it.

For the finite player in us freedom is a function of time. We must have the time to be free. For the infinite player in us time is a function of freedom. We are free to have time. A finite player puts play into time. An infinite player puts time into play.

69

Just as infinite players can play any number of finite games, so too can they join the audience of any game. They do so, however, for the play that is in observing, quite aware that they are audience. They look, but they see that they are looking.

Infinite play remains invisible to the finite observer. Such viewers are looking for closure, for the ways in which players can bring matters to a conclusion and finish whatever remains unfinished. They are looking for the way time has exhausted itself, or will soon do so. Finite players stand before infinite play as they stand before art, looking at it, making a poiema of it.

If, however, the observers see the poiesis in the work they cease at once being observers. They find themselves in its time, aware that it remains unfinished, aware that their reading of the poetry is itself poetry. Infected then by the genius of the artist they recover their own genius, becoming beginners with nothing but possibility ahead of them.

If the goal of finite play is to win titles for their timelessness, and thus eternal life for oneself, the essence of infinite play is the paradoxical engagement with temporality that Meister Eckhart called "eternal birth."

FIVE

Nature Is the Realm of the Unspeakable

70

NATURE IS the realm of the unspeakable. It has no voice of its own, and nothing to say. We experience the unspeakability of nature as its utter indifference to human culture.

The Master Player in us tolerates this indifference scarcely at all. Indeed, we respond to it as a challenge, an invitation to confrontation and struggle. If nature will offer us no home, offer us nothing at all, we will then clear and arrange a space for ourselves. We take nature on as an opponent to be subdued for the sake of civilization. We count among the highest achievements of modern society the development of a technology that allows us to master nature's vagaries.

The effort has largely taken the form of theatricalizing our relation to nature. Like any Master Player we have been patiently attentive to the slightest clues in our opponent's behavior—as a way of preparing ourselves against surprise. Like hunters stalking their prey, we have learned to mimic the movements of nature, waiting for the chance to take hold of them before they get away from us. "Nature, to be commanded, must be obeyed" (Bacon). It is as though, by learning its secret script, we have learned to direct its play as well. There is little left to surprise us.

The assumption guiding our struggle against nature is that

deep within itself nature contains a structure, an order, that is ultimately intelligible to the human understanding. Since this inherent structure determines the way things change, and is not itself subject to change, we speak of nature being lawful, of repeating itself according to quite predictable patterns.

What we have done by showing that certain events repeat themselves according to known laws is to *explain* them. Explanation is the mode of discourse in which we show why matters must be as they are. All laws made use of in explanation look backward in time from the conclusion or the completion of a sequence. It is implicit in all explanatory discourse that just as there is a discoverable necessity in the outcome of past events, there is a discoverable necessity in future events. What can be explained can also be predicted, if one knows the initial events and the laws covering their succession. A prediction is but an explanation in advance.

Because of its thorough lawfulness nature has no genius of its own. On the contrary, it is sometimes thought that the grandest discovery of the human genius is the perfect compatibility between the structure of the natural order and the structure of the mind, thereby making a complete understanding of nature possible. "One may say 'the eternal mystery of the world is its comprehensibility' " (Einstein).

This is as much as to say that nature does have a voice, and its voice is no different from our own. We can then presume to speak for the unspeakable.

This achievement is often raised as a sign of the great superiority of modern civilization over the many faded and lost civilizations of the ancients. While our great skill lies in finding patterns of repetition under the apparent play of accident and chance, less successful civilizations dealt with the threats of natural accident by appealing to supernatural powers for protection. But the voices of the gods proved to be ignorant and false; they have been silenced by the truth.

71

There is an irony in our silencing of the gods. By presuming to speak for the unspeakable, by hearing our own voice as the voice of nature, we have had to step outside the circle of nature. It is one thing for physics and chemistry to be speaking *about* nature; it is quite another for physics and chemistry to be the speaking *of* nature. No chemist would want to say that chemistry is itself chemical, for our speaking cannot be both chemical and about chemistry. If speaking about a process is itself part of the process, there is something that must remain permanently hidden from the speaker. To be intelligible at all, we must claim that we can step aside from the process and comment on it "objectively" and "dispassionately," without anything obstructing our view of these matters. Here lies the irony: By way of this perfectly reasonable claim the gods have stolen back into our struggle with nature. By depriving the gods of their own voices, the gods have taken ours. It is we who speak as supernatural intelligences and powers, masters of the forces of nature.

This irony passes unnoticed only so long as we continue to veil ourselves against what we can otherwise plainly see: nature allows no master over itself. Bacon's principle works both ways. If we must obey to command, then our commanding is only obeying and not commanding at all. There is no such thing as an unnatural act. Nothing can be done to or against nature, much less outside it. Therefore, the ignorance we thought we could avoid by an unclouded observation of nature has swept us back into itself. What we thought we read in nature we discover we have read into nature. "We have to remember that what we observe is not nature in itself but nature exposed to our method of questioning" (Heisenberg).

72

We are speaking now of no ordinary ignorance. It is not what we could have known but do not; it is unintelligibility itself: that which no mind can ever comprehend.

Unveiled, aware of the insuperable limitation placed against all our looking, we come back to nature's perfect silence. Now we can see that it is a silence so complete there is no way of knowing what it is silent about—if anything. What we learn from this silence is the unlikeness between nature and whatever we could think or say about it. But *this* silence has an irony of its own: Far from stupefying us, it provides an indispensable condition to the mind's own originality. By confronting us with radical unlikeness, nature becomes the source of metaphor.

Metaphor is the joining of like to unlike such that one can never become the other. Metaphor requires an irreducibility, an imperturbable indifference of its terms for one another. The falcon can be the "kingdom of daylight's dauphin" only if the daylight could have no dauphin, could indeed have nothing to do with dauphins.

At its root all language has the character of metaphor, because no matter what it intends to be about it remains language, and remains absolutely unlike whatever it is about. This means that we can never have the falcon, only the word "falcon." To say that we have the falcon, and not the "falcon," is to presume again that we know precisely what it is we have, that we can see it in its entirety, and that we can speak as nature itself.

The unspeakability of nature is the very possibility of language.

Our attempt to take control of nature, to be Master Player in our opposition to it, is an attempt to rid ourselves of language. It is the refusal to accept nature as "nature." It is to deafen ourselves to metaphor, and to make nature into something so familiar it is essentially an extension of our willing and speaking. What the hunter kills is not the deer, but the metaphor of the deer—the "deer." Killing the deer is not an act against nature; it is an act against language. To kill is to impose a silence that remains a silence. It is the reduction of an unpredictable vitality to a predictable mass, the transformation of the remote into the familiar. It is to rid oneself of the need to attend to its otherness.

The physicists who look at their objects within their limitations teach physics; those who see the limitations they place around their objects teach "physics." For them physics is a poiesis.

73

If nature is the realm of the unspeakable, history is the realm of the speakable. Indeed, no speaking is possible that is not itself historical. Students of history, like students of nature, often believe they can find unbiased, direct views of events. They look in on the lives of others, noting the multitude of ways those lives have been limited by the age in which they were lived. But no one can look in on an age, even if it is one's own age, without looking out of an age as well. There is no refuge outside history for such viewers, any more than there is a vantage outside nature.

Since history is the drama of genius, its relentless surprise tempts us into designing boundaries for it, searching through it for patterns of repetition. Historians sometimes speak of trends, of cycles, of currents, of forces, as though they were describing natural events. In doing so they must dehistoricize themselves, taking a perspective from the timeless, believing that each observed history is always of others and never of themselves, that each observation is of history but not itself historical.

Genuine historians therefore reverse the assumption of the observers of nature that the observation itself cannot be an act of nature. Historians who understand themselves to be historical abandon explanation altogether. The mode of discourse appropriate to such self-aware history is *narrative*.

Like explanation, narrative is concerned with a sequence of events and brings its tale to a conclusion. However, there is no general law that makes this outcome necessary. In a genuine story there is no law that makes *any* act necessary. Explanations place all apparent possibilities into the context of the necessary; stories set all necessities into the context of the possible.

Explanation can tolerate a degree of chance, but it cannot comprehend freedom at all. We explain nothing when we say that persons do whatever they do because they choose to do it. On the other hand, causation cannot find a place in narrative. We have not told a story when we show that persons do whatever they do because they were caused to do it—by their genes, their social circumstances, or the influence of the gods.

Explanations settle issues, showing that matters must end as they have. Narratives raise issues, showing that matters do not end as they must but as they do. Explanation sets the need for further inquiry aside; narrative invites us to rethink what we thought we knew.

If the silence of nature is the possibility of language, language is the possibility of history.

74

Successful explanations do not draw attention to themselves as modes of speaking, because what is explained is not itself subject to history. If I explain to you why cold water sinks to the bottom of the pond and ice rises to the surface, I certainly do not intend my explanation to be true now and not later. The explanation is true anywhere and any time.

That I choose to explain this to you in this time and place, however, *is* historical; it is an event—the narrative of our relation with each other. There must therefore be a reason for the speaking of this verity. Explanations are not offered gratuitously, just because, say, ice happens to float. I can explain nothing to you unless I first draw your attention to patent inadequacies in your knowledge: discontinuities in the relations between objects, or the presence of anomalies you cannot account for by any of the laws known to you. You will remain deaf to my explanations until you suspect yourself of falsehood.

Many of these suspicions are, of course, minor, requiring merely small adjustments in one's views, incurring no doubt whatsoever concerning those views. Major challenges, however, are too serious to be met with argument, or with sharpened explanation. They call either for outright and wholesale rejection, or for conversion. One does not cross over from Manichaeism to Christianity, or from Lamarckianism to Darwinism, by a mere adjustment of views. True conversions consist in the choice of a new audience, that is, of a new world. All that was once familiar is now seen in startlingly new ways.

As theatrical as conversions are, they remain oblivious to the degree to which choice is involved in the passage from one world to another. Radical conversions, especially, veil themselves against their own arbitrariness. Augustine, the most famous convert of antiquity, was puzzled that he could have held so firmly to so many different falsehoods; he was not astounded that there are so many different truths. His conversion was not from explanation to narrative, but from one explanation to another. When he crossed the line from paganism to Christianity, he arrived in the territory of a truth beyond further challenge.

Explanations succeed only by convincing resistant hearers of their error. If you will not hear my explanations until you are suspicious of your own truths, you will not accept my explanations until you are convinced of your error. Explanation is an antagonistic encounter that succeeds by defeating an opponent. It possesses the same dynamic of resentment found in other finite play. I will press my explanations on you because I need to show that I do not live in the error that I think others think I do.

Whoever wins this struggle is privileged with the claim to true knowledge. Knowledge has been *arrived at*, it is the outcome of this engagement. Its winners have the uncontested power to make certain statements of fact. They are to be listened to. In those areas appropriate to the contests now concluded, winners possess a knowledge that no longer can be challenged.

Knowledge, therefore, is like property. It must be published, declared, or in some other way so displayed that others cannot but take account of it. It must stand in their way. It must be emblematic, pointing backward at its possessor's competitive skill.

So close are knowledge and property that they are often thought to be continuous. Those who are entitled to knowledge feel they should be granted property as well, and those who are entitled to property believe a certain knowledge goes with it. Scholars demand higher salaries for their publishable successes; industrialists sit on university boards.

75

If explanation, to be successful, must be oblivious to the silence of nature, it must also in its success impose silence on its listeners. Imposed silence is the first consequence of the Master Player's triumph.

What one wins in a title is the privilege of magisterial speech. The privilege of magisterial speech is the highest honor attaching to any title. We expect the first act of a winner to be a speech. The first act of the loser may also be a speech, but it will be a speech to concede victory, to declare there will be no further challenge to the winner. It is a speech that promises to silence the loser's voice.

The silence to which the losers pledge themselves is the silence of obedience. Losers have nothing to say; nor have they an audience who would listen. The vanquished are effectively of one will with the victors, and of one mind; they are completely incapable of opposition, and therefore without any otherness whatsoever.

The victorious do not speak with the defeated; they speak for the defeated. Husbands speak for wives in the finite family, and parents for their children. Kings speak for the realm, governors for the state, popes for the church. Indeed, the titled, as titled, cannot speak *with* anyone.

It is chiefly in magisterial speech that the power of winners resides. To be powerful is to have one's words obeyed. It is only by magisterial speech that the emblematic property of winners can be safeguarded. Those entitled to their possessions have the privilege of *calling* the police, *calling up* an army, to force the recognition of their emblems.

The power of gods is known principally through their utterances. The *sicut dixit dominus* (thus says the lord) is always a signal for ritual silence. The speech of a god can be so perfectly expressive of that god's power that the god and its speech become identical: "In the beginning was the word. The word was with God, and the word was God."

One is speechless before a god, or silent before a winner, because it no longer matters to others what one has to say. To lose a contest is to become obedient; to become obedient is to lose one's listeners. The silence of obedience is an unheard silence. It is the silence of death. For this reason the demand for obedience is inherently evil.

The silence of nature is the possibility of language. By subduing nature the gods give it their own voice, but in making nature an opponent they make all their listeners opponents. By refusing the silence of nature they demand the silence of obedience. The unspeakability of nature is therefore transformed into the unspeakability of language itself.

76

Infinite speech is that mode of discourse that consistently reminds us of the unspeakability of nature. It bears no claim to truth, originating from nothing but the genius of the speaker. Infinite speech is therefore not *about* anything; it is always *to* someone. It is not command, but address. It belongs entirely to the speakable.

That language is not about anything gives it its status as metaphor. Metaphor does not point at something there. Never shall we find the kingdom of daylight's dauphin in one place or another. It is not the role of metaphor to draw our sight to what is there, but to draw our vision toward what is not there and, indeed, cannot be anywhere. Metaphor is horizonal, reminding us that it is one's vision that is limited, and not what one is viewing.

The meaning of a finite speaker's discourse lies in what precedes its utterance, what is already the case and therefore is the case *whether or not* it is spoken.

The meaning of an infinite speaker's discourse lies in what comes of its utterance—that is, whatever is the case *because it is spoken.*

Finite language exists complete before it is spoken. There is first a language—*then* we learn to speak it. Infinite language exists only as it is spoken. There is first a language—*when* we learn to speak it. It is in this sense that infinite discourse always arises from a perfect silence.

Finite speakers come to speech with their voices already trained and rehearsed. They must know what they are doing with the language before they can speak it. Infinite speakers must wait to see what is done with their language by the listeners before they can know what they have said. Infinite speech does not expect the hearer to see what is already known to the speaker, but to share a vision the speaker could not have had without the response of the listener.

Speaker and listener understand each other not because they have the same knowledge about something, and not be-

cause they have established a likeness of mind, but because
they know "how to go on" with each other (Wittgenstein).

77

Because it is address, attending always on the response of
the addressed, infinite speech has the form of listening. Infinite
speech does not end in the obedient silence of the hearer,
but continues by way of the attentive silence of the speaker.
It is not a silence into which speech has died, but a silence
from which speech is born.

Infinite speakers do not give voice to another, but receive
it from another. Infinite speakers do not therefore appeal to
a world as audience, do not speak before a world, but present
themselves as an audience by way of talking with others. Finite
speech informs another about the world—for the sake of being
heard. Infinite speech forms a world about the other—for
the sake of listening.

It is for this reason that the gods, insofar as they speak as
the lords of this world, magisterially, speak before this world
and are therefore unable to change it. Such gods cannot create
a world but can only be creations of a world—can only be
idols. A god cannot create a world and be magisterial within
it. "The religions which represent divinity as commanding
whenever it has the power to do so seem false. Even though
they are monotheistic they are idolatrous" (Weil).

A god can create a world only by listening.

Were the gods to address us it would not be to bring us
to silence through their speech, but to bring us to speech
through their silence.

The contradiction of finite speech is that it must end by
being heard. The paradox of infinite speech is that it continues

only because it is a way of listening. Finite speech ends with a silence of closure. Infinite speech begins with a disclosure of silence.

78

Storytellers do not convert their listeners; they do not move them into the territory of a superior truth. Ignoring the issue of truth and falsehood altogether, they offer only vision. Storytelling is therefore not combative; it does not succeed or fail. A story cannot be obeyed. Instead of placing one body of knowledge against another, storytellers invite us to return from knowledge to thinking, from a bounded way of looking to an horizonal way of seeing.

Infinite speakers are Plato's poietai taking their place in the historical. Storytellers enter the historical not when their speaking is full of anecdotes about actual persons, or when they appear as characters in their own tales, but when in their speaking we begin to see the narrative character of *our* lives. The stories they tell touch us. What we thought was an accidental sequence of experiences suddenly takes the dramatic shape of unresolved narrative.

There is no narrative without structure, or plot. In a great story this structure seems like fate, like an inescapable judgment descending on its still unaware heroes, a great metaphysical causality that crowds out all room for choice. Fate arises not as a limitation on our freedom, but as a manifestation of our freedom, testimony that choice is consequent. The exercise of your freedom cannot prevent the exercise of my own freedom, but it can determine the context in which I am to act freely. You cannot make choices for me, but you can largely determine what my choices will be about. Great stories explore the drama of this deeper touching of one free person by another. They are therefore genuinely sexual dramas astounding us once more with the magic of origins.

The myth of Oedipus is one of many great sexual narratives in the cultural treasury of the West that plays on the dramatic relation of fate and origin. Once Oedipus had impulsively killed Laius, not knowing he was his own father, he was carried ahead by ambition and lust into marriage with the dead man's wife, unaware of her true identity. We can read this as fate or we can read it as one act of willfulness that leads to another. Oedipus has taken the posture of the Master Player executing terminal moves—but the moves are not terminal. Oedipus is able to bring nothing to an end. Even the act of blinding himself, meant as a kind of concluding gesture, only brings him to a higher vision. What Oedipus sees is not what the gods have done to him, but what he has done. He learns that what had been limited was his vision, and not what he was viewing. His blinding is an unveiling, and like all unveiling it is self-unveiling. Confronted in the end with nothing but his own genius, Oedipus is finally able to touch. The end of his story is a beginning.

What raises this story into the historical is not just that Oedipus sees; it is that we see that he sees. We become listeners who see that we are listening and therefore participating in a now enlarged drama of origins. Nothing is explained here. On the contrary, what we see is that everything remains still to be said.

79

There is a risk here of supposing that because we know our lives to have the character of narrative, we also know what that narrative is. If I were to know the full story of my life I would then have translated it back into explanation. It is as though I could stand as audience to myself, seeing the opening scene and the final scene at the same time, as though I could see my life in its entirety. In doing so I would be performing it, not living it.

Societal theorists are tempted into the belief that they know the story of a civilization. They can script its final scene of triumph or defeat. It is by way of such end-of-history thinking that the discovered laws of behavior to which persons conform become the scripted laws of behavior to which they must conform.

True storytellers do not know their own story. What they listen to in their poiesis is the disclosure that wherever there is closure there is the possibility of a new opening, that they do not die at the end, but in the course of play. Neither do they know anyone else's story in its entirety. The primary work of historians is to open all cultural termini, to reveal continuity where we have assumed something has ended, to remind us that no one's life, and no culture, can be *known*, as one would know a poiema, but only *learned*, as one would learn a poiesis.

Historians become infinite speakers when they see that whatever begins in freedom cannot end in necessity.

SIX

WE CONTROL NATURE FOR SOCIETAL REASONS

80

WE CONTROL NATURE for societal reasons. The control of nature advances with our ability to predict the outcome of natural processes. Inasmuch as predictions are but explanations in reverse, it is possible that they will be quite as combative as explanations. Indeed, prediction is the most highly developed skill of the Master Player, for without it control of an opponent is all the more difficult. It follows that our domination of nature is meant to achieve not certain natural outcomes, but certain societal outcomes.

A small group of physicists, using calculations of the highest known abstraction, uncovered a predictable sequence of sub-atomic reactions that led directly to the construction of a thermonuclear bomb. It is true that the successful detonation of the bomb proved the predictions of the physicists, but it is also true that we did not explode the bomb to prove them correct; we exploded it to control the behavior of millions of persons and to bring our relations with them to a certain closure.

What this example shows is not that we can exercise power over nature, but that our attempt to do so masks our desire for power over each other. This raises a question as to the *cultural* consequences of abandoning the strategy of power in our attitude toward nature.

The alternative attitudes toward nature can be characterized in a rough way by saying that the result of approaching nature as a hostile Other whose designs are basically inimical to our interests is the *machine*, while the result of learning to discipline ourselves to consist with the deepest discernable patterns of natural order is the *garden*.

"Machine" is used here as inclusive of technology and not as an example of it—as a way of drawing attention to the mechanical rationality of technology. We might be surprised by the technological devices that spring from the imagination of gifted inventors and engineers, but there is nothing surprising in the technology itself. The physicists' bomb is as thoroughly mechanical as the Neanderthal's lever—each the exercise of calculable cause-and-effect sequences.

"Garden" does not refer to the bounded plot at the edge of the house or the margin of the city. This is not a garden one lives beside, but a garden one lives within. It is a place of growth, of maximized spontaneity. To garden is not to engage in a hobby or an amusement; it is to design a culture capable of adjusting to the widest possible range of surprise in nature. Gardeners are acutely attentive to the deep patterns of natural order, but are also aware that there will always be much lying beyond their vision. Gardening is a horizonal activity.

Machine and garden are not absolutely opposed to each other. Machinery can exist in the garden quite as finite games can be played within an infinite game. The question is not one of restricting machines from the garden but asking whether a machine serves the interest of the garden, or the garden the interest of the machine. We are familiar with a kind of mechanized gardening that has the appearance of high productivity, but looking closely we can see that what is intended is not the encouragement of natural spontaneity but its harnessing.

81

The most elemental difference between the machine and the garden is that one is driven by a force which must be introduced from without, the other grown by an energy which originates from within itself.

Certainly machines of extraordinary complexity have been built: spacecraft, for example, that sustain themselves for months in the void while performing complicated functions with great accuracy. But no machine has been made, nor can one be made, that has the source of its spontaneity within itself. A machine must be designed, constructed, and fueled.

Certainly gardens can be treated with such a range of chemical and technological strategies that we can speak of "raising" food, and of the food we have raised as "produce." But no way has been found, or can be found, by which organic growth can be forced from without. The application of fertilizers, herbicides, and any number of other substances does not *alter* growth but *allows* growth; it is meant to consist with natural growth. A plant cannot be designed or constructed. Though we seem to give it "fuel" in the form of rich earth and appropriate nutrients, we depend on the plant to make use of the fuel by way of its own vitality. A machine depends on its designer and its operator both for the supply of fuel and its consumption. A machine has not the merest trace of its own spontaneity or vitality. Vitality cannot be given, only found.

82

Just as nature has no outside, it has no inside. It is not divided within itself and cannot therefore be used for or against itself. There is no inherent opposition of the living and the nonliving within nature; neither is more or less natural than the other. The use of agricultural poisons, for example, will surely kill selected organisms; it will arrest the spontaneity of living entities—but it is not an unnatural act. Nature has not been changed. All that changes is the way we discipline ourselves to consist with natural order.

Our freedom in relation to nature is not the freedom to change nature; it is not the possession of power over natural phenomena. It is the freedom to change ourselves. We are perfectly free to design a culture that will turn on the awareness that vitality cannot be given but only found, that the given patterns of spontaneity in nature are not only to be respected, but to be celebrated.

Although "natural order" is the common expression, it has something of a veiling quality about it. More properly speaking, it is not the order of nature but its irreducible spontaneity with which we find ourselves contending. That nature has no outside, and no inside, that it suffers no opposition to itself, that it is not moved by unnatural influence, is not the expression of an order so much as it is the display of a perfect *indifference* on nature's part to all matters cultural.

Nature's source of movement is always from within itself; indeed it *is* itself. And it is radically distinct from our own source of movement. This is not to say that, possessing no order, nature is chaotic. It is neither chaotic nor ordered. Chaos and order describe the cultural experience of nature— the degree to which nature's indifferent spontaneity seems to agree with our current manner of cultural self-control. A hurricane, or a plague, or the overpopulation of the earth will seem chaotic to those whose cultural expectations are damaged by them and orderly to those whose expectations have been confirmed by them.

83

The *paradox* in our relation to nature is that the more deeply a culture respects the indifference of nature, the more creatively it will call upon its own spontaneity in response. The more clearly we remind ourselves that we can have no unnatural influence on nature, the more our culture will embody a freedom to embrace surprise and unpredictability.

Human freedom is not a freedom over nature; it is the freedom to be natural, that is, to answer to the spontaneity of nature with our own spontaneity. Though we are free to be natural, we are not free by nature; we are free by culture, by history.

The *contradiction* in our relation to nature is that the more vigorously we attempt to force its agreement with our own designs the more subject we are to its indifference, the more vulnerable to its unseeing forces. The more power we exercise over natural process the more powerless we become before it. In a matter of months we can cut down a rain forest that took tens of thousands of years to grow, but we are helpless in repulsing the desert that takes its place. And the desert, of course, is no less natural than the forest.

Such contradiction is most obvious in the matter of machinery. We make use of machines to increase our power, and therefore our control, over natural phenomena. By exerting themselves no more than is necessary to operate fingertip controls, a team of workers can cut six-lane highways through mountains and dense forest, or fill in wetlands to build shopping malls.

While a machine greatly aids the operator in such tasks, it also disciplines its operator. As the machine might be considered the extended arms and legs of the worker, the worker might be considered an extension of the machine. All machines, and especially very complicated machines, require operators to place themselves in a provided location and to perform functions mechanically adapted to the functions of the machine. To use the machine for control is to be controlled by the machine.

To operate a machine one must operate like a machine. Using a machine to do what we cannot do, we find we must do what the machine does.

Machines do not, of course, *make* us into machines when we operate them; we make ourselves into machinery in order to operate them. Machinery does not steal our spontaneity from us; we set it aside ourselves, we deny our originality. There is no *style* in operating a machine. The more efficient the machine, the more it either limits or absorbs our uniqueness into its operation.

Indeed, we come to think that the style of operation does not belong to the operator at all, but is inherent in the machine. Advertisers and manufacturers speak of their products as though they have designed style into them. Most consumer products are "styled" inasmuch as they actually standardize the activity or the taste of the consumer. In a perfect contradiction we are urged to buy a "styled" artifact *because* others are also buying it—that is, we are asked to express our genius by giving up our genius.

Because we make use of machinery in the belief we can increase the range of our freedom, and instead only decrease it, *we use machines against ourselves.*

85

Machinery is contradictory in another way. Just as we use machinery against ourselves, *we also use machinery against itself.* A machine is not a way of doing something; it stands in the way of doing something.

When we use machines to achieve whatever it is we desire, we cannot have what we desire until we have finished with the machine, until we can rid ourselves of the mechanical means of reaching our intended outcome. The goal of technology is therefore to eliminate itself, to become silent, invisible, carefree.

We do not purchase an automobile, for example, merely to own some machinery. Indeed, it is not *machinery* we are buying at all, but what we can have by way of it: a means of rapidly carrying us from one location to another, an object of envy for others, protection from the weather. Similarly, a radio must cease to exist as equipment and become sound. A perfect radio will draw no attention to itself, will make it seem we are in the very presence of the source of its sound. Neither do we watch a movie screen, nor look at television. We look at what is *on* television, or *in* the movie, and become annoyed when the equipment intrudes—when the film is unfocused or the picture tube malfunctions.

When machinery functions perfectly it ceases to be there—but so do we. Radios and films allow us to be where we are not and not be where we are. Morever, machinery is veiling. It is a way of hiding our inaction from ourselves under what appear to be actions of great effectiveness. We persuade ourselves that, comfortably seated behind the wheels of our autos, shielded from every unpleasant change of weather, and raising or lowering our foot an inch or two, we have actually traveled somewhere.

Such travel is not *through* space foreign to us, but *in* a space that belongs to us. We do not move from our point of departure, but with our point of departure. To be moved from our living room by an automobile whose upholstered seats differ scarcely at all from those in our living rooms, to an airport waiting room and then to the airplane where we are provided the same sort of furniture, is to have taken our origin with us; it is to have left home without leaving home. To be at home everywhere is to neutralize space.

Therefore, the importance of reducing time in travel: by arriving as quickly as possible we need not feel as though we had left at all, that neither space nor time can affect us—as though they belong to us, and not we to them.

We do not go somewhere in a car, but arrive somewhere in a car. Automobiles do not make travel possible, but make it possible for us to move locations without traveling.

Thus, the theatricality of machinery: Such movement is but a change of scenes. If effective, the machinery will see to it that we remain untouched by the elements, by other travelers, by those whose towns or lives we are traveling through. We can see without being seen, move without being touched.

When most effective, the technology of communication allows us to bring the histories and the experiences of others into our home, but without changing our home. When most effective, the technology of travel allows us to pass through the histories of other persons with the "comforts of home," but without changing those histories.

When it is most effective, machinery will have no effect at all.

86

In still another way is machinery contradictory. Using it against itself and against ourselves, *we also use machinery against each other.*

I cannot use machinery without using it with another. I do not talk on the telephone; I talk with someone on the telephone. I listen to someone on the radio, drive to visit a friend, compute business transactions. To the degree that my association with you depends on such machinery, the connecting medium makes each of us an extension of itself. If your business activities cannot translate into data recognizable by my computer, I can have no business with you. If you do not live where I can drive to see you, I will find another friend. In each case your relationship to me does not depend on my needs but on the needs of my machinery.

If to operate a machine is to operate like a machine, then *we not only operate with each other like machines, we operate each other like machines.* And if a machine is most effective when it has no effect, then we operate each other in such a way that we reach the outcome desired—in such a way that nothing happens.

The inherent hostility of machine-mediated relatedness is nowhere more evident than in the use of the most theatrical machines of all: instruments of war. All weapons are designed to affect others without affecting ourselves, to make others answerable to the technology in our control. Weapons are the equipment of finite games designed in such a way that they do not maximize the play but eliminate it. Weapons are meant not to win contests but to end them. Killers are not victors; they are unopposed competitors, players without a game, living contradictions.

This is particularly the case with the airborne electronic weaponry of the present century, where the operator deals only with the technology—buttons, blips, lights, dials, levers, computer data—and never with the unseen opponent. Indeed, so empty of drama is the modern machinery of slaughter that it is intended to assault enemies only while they are still unseen. This reaches an extreme form in the belief that our enemies are not unseen because they are enemies, but are enemies because they are unseen.

There is a logic in the instrumentality of death that leads us to killing the unseen because they are unseen. The crudest spear or sword is raised by an attacker because the independent existence of another cannot be countenanced—*because the other cannot be seen as an other.* Just as I insist that the condition of our friendship is your unresisting use of the telephone, I will expect the weapon in my hand to function without finding an other that can resist it. Killers can suffer no suggestion that they are living into the open, that their histories are not finished, that their freedom is always a freedom with others, and not over others, that it is not their vision that is limited but what they are viewing.

The fact that the technology of slaughter at vast distances has become extremely sophisticated does not culturally advance its highly trained operators over club-swinging primitives; it makes complete the blindness that was but rudimentary in the primitive. It is the supreme triumph of resentment over vision. We are the unseeing killing the unseen.

Not everyone who uses machinery is a killer. But when the use of machinery springs from our attempt to respond to the indifference of nature with an indifference of our own to nature, we have begun to acquire the very indifference to persons that has led to the century's grandest crimes by its most civilized nations.

87

If indifference *to* nature leads to the machine, the indifference *of* nature leads to the garden. All culture has the form of gardening: the encouragement of spontaneity in others by way of one's own, the respect for source, and the refusal to convert source into resource.

Gardeners slaughter no animals. They kill nothing. Fruits, seeds, vegetables, nuts, grains, grasses, roots, flowers, herbs, berries—all are collected when they have ripened, and when their collection is in the interest of the garden's heightened and continued vitality. Harvesting respects a source, leaves it unexploited, suffers it to be as it is.

Animals cannot be harvested. They mature, but they do not "ripen." They are killed not when they have completed the cycle of their vitality but when they are at the peak of their vitality. Finite gardeners, converting agriculture into commerce, "raise" or "produce" animals—or meat products—as though by machine. Animal husbandry is a science, a method of controlling growth. It assumes that animals belong to us. What is source in them is to be resource for us. Cattle are confined to pens to prevent such movement as would "toughen" their flesh. Geese, their feet nailed to the floor, are force fed like machines until they can be butchered for their fattened livers.

While machinery is meant to work changes without changing its operators, gardening transforms its workers. One learns how to drive a car, one learns to drive as a car; but one becomes a gardener.

Gardening is not outcome-oriented. A successful harvest is not the end of a garden's existence, but only a phase of it. As any gardener knows, the vitality of a garden does not end with a harvest. It simply takes another form. Gardens do not "die" in the winter but quietly prepare for another season.

Gardeners celebrate variety, unlikeness, spontaneity. They understand that an abundance of styles is in the interest of vitality. The more complex the organic content of the soil, for example—that is, the more numerous its sources of change—the more vigorous its liveliness. Growth promotes growth.

So also in culture. Infinite players understand that the vigor of a culture has to do with the variety of its sources, the differences within itself. The unique and the surprising are not suppressed in some persons for the strength of others. The genius in you stimulates the genius in me.

One operates a machine effectively, so that it disappears, giving way to results in which the machine has no part. One gardens creatively, so that all the sources of the garden's vitality appear in its harvest, giving rise to a continuity in which we take an active part.

Inasmuch as gardens do not conclude with a harvest and are not played for a certain outcome, one never arrives anywhere with a garden.

A garden is a place where growth is found. It has its own source of change. One does not bring change to a garden, but comes to a garden prepared for change, and therefore prepared to change. It is possible to deal with growth only out of growth. True parents do not see to it that their children grow in a particular way, according to a preferred pattern or scripted stages, but they see to it that they grow with their children. The character of one's parenting, if it is genuinely dramatic, must be constantly altered from within as the children change from within. So, too, with teaching, or working with, or loving each other.

It is in the garden that we discover what travel truly is. We do not journey to a garden but by way of it.

Genuine travel has no destination. Travelers do not go somewhere, but constantly discover they are somewhere else. Since gardening is a way not of subduing the indifference of nature but of raising one's own spontaneity to respond to the disregarding vagaries and unpredictabilities of nature, we do not look on nature as a sequence of changing scenes but look on ourselves as persons in passage.

Nature does not change; it has no inside or outside. It is therefore not possible to travel *through* it. All travel is therefore change within the traveler, and it is for that reason that travelers are always somewhere else. To travel is to grow.

Genuine travelers travel not to overcome distance but to discover distance. It is not distance that makes travel necessary, but travel that makes distance possible. Distance is not determined by the measurable length between objects, but by the actual differences between them. The motels around the airports in Chicago and Atlanta are so little different from the motels around the airports of Tokyo and Frankfurt that all essential distances dissolve in likeness. What is truly separated is distinct; it is unlike. "The only true voyage would be not to travel through a hundred different lands with the same pair of eyes, but to see the same land through a hundred different pairs of eyes" (Proust).

A gardener, whose attention is ever on the spontaneities of nature, acquires the gift of seeing differences, looks always for the merest changes in plant growth, or in the composition of the soil, the emerging populations of insects and earthworms. So will gardeners, as parents, see changes of the smallest subtlety in their children, or as teachers see the signs of an increasing skill, and possibly wisdom, in their students. A garden, a family, a classroom—any place of human gathering whatsoever—will offer no end of variations to be observed, each an arrow pointing toward yet more changes. But these observed changes are not theatrically amusing to genuine gardeners; they dramatically open themselves to a renewed future.

So, too, with those who look everywhere for difference, who see the earth as source, who celebrate the genius in others, who are not prepared against but for surprise. "I have traveled far in Concord" (Thoreau).

89

Since machinery requires force from without, its use always requires a search for consumable power. When we think of nature as resource, it is as a resource for power. As we preoccupy ourselves with machinery, nature is increasingly thought of as a reservoir of needed substances. It is a quantity of materials that exist to be consumed, chiefly in our machines.

Being undivided, nature cannot be used against itself. We do not therefore consume *it*, or exhaust *it*. We simply rearrange our societal patterns in a way that reduces our ability to respond creatively to the existing patterns of spontaneity. That is, to use the societal expression, we create *waste*. Waste, of course, is by no means unnatural. The trash and garbage of a civilization do not befoul nature; they are nature—but in a form society no longer is able to exploit for its own ends.

Society regards its waste as an unfortunate, but necessary, consequence of its activities—what is left when we have made essential societal goods available. But waste is not the *result* of what we have made. It *is* what we have made. Waste plutonium is not an indirect consequence of the nuclear industry; it is a product of that industry.

90

Waste is unveiling. As we find ourselves standing in garbage that we know is our own, we find also that it is garbage we have *chosen* to make, and having chosen to make it could choose not to make it. Because waste is unveiling, we remove it. We place it where it is out of sight. We either find uninhabited areas where waste can be disposed of, or fill them with our refuse until they become uninhabitable. Since a flourishing society will vigorously exploit its natural resources, it will produce correspondingly great quantities of trash, and quickly its uninhabited lands will overflow with waste, threatening to make the society's own habitation into a wasteland.

Because waste is unveiling, it is not only placed out of sight, it is declared a kind of antiproperty. No one owns it. Part of the contradiction in the phenomenon of waste is that treating nature as though it belongs to us we must soon treat nature as though it belongs to no one. Not only does no one own waste, no one wants it. Instead of competing to possess these particular products, we compete to dispossess them. We force it on others less able to rid themselves of it. Trash accumulates in slums, sewage runs downstream, airborne acid drifts hundreds of miles settling on the lands of those powerless to halt its "disposal" into the atmosphere. Thousands of square miles of farm lands have been laid waste by the construction of multilane highways, or submerged by dams whose water is used to flush waste from distant cities.

Waste is the antiproperty that becomes the possession of losers. It is the emblem of the untitled.

Waste is unveiling, because it persists in showing itself *as* waste, and as *our* waste. If waste is the result of our indifference to nature, it is also the way we experience the indifference of nature. Waste is therefore a reminder that society is a species of culture. Looking about at the wasteland into which we have converted our habitation, we can plainly see that nature is not whatever we want it to be; but we can also plainly see that society is only what we want it to be.

It is a consequence of this contradiction that the more waste a society produces, the more unveiling that waste is, and thus the more vigorously must a society deny that it produces any waste at all; the more it must dispose, or hide, or ignore, its detritus.

Since the attempt to control nature is at its heart the attempt to control other persons, we can expect societies to be less patient with those cultures which express some degree of indifference to societal goals and values. It is this repeated parallel that brings us to see that the society that creates natural waste creates human waste.

Waste persons are those no longer useful as resources to a society for whatever reason, and have become *apatrides*, or noncitizens. Waste persons must be placed out of view—in ghettos, slums, reservations, camps, retirement villages, mass graves, remote territories, strategic hamlets—all places of desolation, and uninhabitable. We live in a century whose Master Players have created many millions of such "superfluous persons" (Rubenstein).

A people does not become superfluous by itself, any more than natural waste creates itself. It is society that declares some persons to be waste. Human trash is not an unfortunate burden on a society, an indirect result of its proper conduct; it is its direct product. European settlers in the American, African, and Asian continents did not happen to come upon populations of unwanted persons nature had thrust in their way; they made them superfluous by way of some of the most important and irreversible principles of their societies.

Strictly speaking, waste persons do not exist outside the boundaries of a society. They are not society's enemies. One does not go to war against them, as one goes to war against another society. Waste persons do not constitute an alternative or threatening society; they constitute an unveiling culture. They are therefore "purged." A society cleanses itself of them.

92

When society is unveiled, when we see that it is whatever we want it to be, that it is a species of culture with nothing necessary in it, by no means a phenomenon of nature or a manifestation of instinct, nature is no longer shaped and fitted into one or another set of societal goals. Unveiled, we stand before a nature whose only face is its hidden self-origination: its genius.

We see nature as genius when we see as genius.

We understand nature as source when we understand ourselves as source. We abandon all attempts at an explanation of nature when we see that we cannot be explained, when our own self-origination cannot be stated as fact. We behold the irreducible otherness of nature when we behold ourselves as its other.

93

For the infinite player, seeing as genius, nature is the absolutely unlike. The infinite player recognizes nothing on the face of nature. Nature displays not only its indifference to human existence but its difference as well.

Nature offers no home. Although we become gardeners in response to its indifference, nature does nothing of itself to feed us. In Jewish and Islamic mythology God provided us with a garden but did not, indeed could not, do the gardening for us. It was only a garden because we could respond to it, because we could be responsible for it. Our responsibility lay in noting its variabilities and discrete features. We were to name the animals, separating one from the other. This garden was not a machine-like device automatically providing food for us. Neither were we machine-like, driven from without and destined from within. According to this myth, God did breathe life into us, but in order to continue living we had to do our own breathing.

But responsibility for the garden does not mean that we can *make* a garden of nature, as though it were a poiema of which we could take possession. A garden is not something we have, over which we stand as gods. A garden is a poiesis, a receptivity to variety, a vision of differences that leads always to a making of differences. The poet joyously suffers the unlike, reduces nothing, explains nothing, possesses nothing.

We stand before genius in silence. We cannot speak it, we can only speak as it. Yet, though I speak as genius, I cannot speak for genius. I cannot give nature a voice in my script. I can not give others a voice in my script—without denying their own source, their originality. To do so is to cease responding to the other, to cease being responsible. No one and nothing *belong* in my script.

The homelessness of nature, its utter indifference to human existence, disclose to the infinite player that nature is the genius of the dramatic.

SEVEN

MYTH PROVOKES EXPLANATION BUT ACCEPTS NONE OF IT

94

MYTH PROVOKES explanation but accepts none of it. Where explanation absorbs the unspeakable into the speakable, myth reintroduces the silence that makes original discourse possible.

Explanations establish islands, even continents, of order and predictability. But these regions were first charted by adventurers whose lives are narratives of exploration and risk. They found them only by mythic journeys into the wayless open. When the less adventuresome settlers arrive later to work out the details and domesticate these spaces, they easily lose the sense that all this firm knowledge does not expunge myth, but floats in it.

Few discoveries were greater than Copernicus', for they projected an order into the heavens that no one has successfully challenged. Many thought then, and some still think, that this great statement of truth dispelled clouds of myth that had kept humankind in retarding darkness. What Copernicus dispelled, however, were not myths but other explanations. Myths lie elsewhere. To see where, we do not look at the facts in Copernicus' works; we look for the story in his stating them. Knowledge is what successful explanation has led to; the thinking that sent us forth, however, is pure story.

Copernicus was a traveler who went with a hundred pairs

of eyes, daring to look again at all that is familiar in the hope of vision. What we hear in this account is the ancient saga of the solitary wanderer, the *peregrinus,* who risks anything for the sake of surprise. True, at a certain point he stopped to look and may have ended his journey as a Master Player setting down bounded fact. But what resounds most deeply in the life of Copernicus is the journey that made knowledge possible and not the knowledge that made the journey successful.

That myth does not accept the explanations it provokes we can see in the boldness with which thinkers in any territorial endeavor reexamine the familiar for a higher seeing. Indeed, the very liveliness of a culture is determined not by how frequently these thinkers discover new continents of knowledge but by how frequently they depart to seek them.

A culture can be no stronger than its strongest myths.

95

A story attains the status of myth when it is retold, and persistently retold, solely for its own sake.

If I tell a story as a way of bracing up an argument or amusing an audience, I am not telling it for its own sake. To tell a story for its own sake is to tell it for no other reason than that it is a story. Great stories have this feature: To listen to them and learn them is to become their narrators.

Our first response to hearing a story is the desire to tell it ourselves—the greater the story the greater the desire. We will go to considerable time and inconvenience to arrange a situation for its retelling. It is as though the story is itself seeking the occasion for its recurrence, making use of us as its agents. We do not go out searching for stories for ourselves; it is rather the stories that have found us for themselves.

Great stories cannot be observed, any more than an infinite game can have an audience. Once I hear the story I enter into its own dimensionality. I inhabit its space at its time. I do not therefore understand the story in terms of my experience, but my experience in terms of the story. Stories that have the enduring strength of myths reach through experience to touch the genius in each of us. But experience is the result of this generative touch, not its cause. So far is this the case that we can even say that *if we cannot tell a story about what happened to us, nothing has happened to us.*

It was not Freud's theory of the unconscious that led him to Oedipus, but the myth of Oedipus that shaped the way he listened to his patients. "The theory of instincts," he wrote, "is so to say our mythology." So too, then, the theory of the unconscious that follows from it, and the superego, and the ego. This is a mythology of such poetic strength that it has altered not only the way we understand our experience, but our experience itself. Who of us has not known a crisis of ego, the disturbing presence of unwanted feelings, or the anxious recoil from a more polymorphously embodied sexuality? These experiences are not described by Freud the dispassionate scientist; they are made possible by Freud the mythic dreamer.

As myths make individual experience possible, they also make collective experience possible. Whole civilizations rise from stories—and can rise from nothing else. It is not the historical experience of Jews that makes the Torah meaningful. The Torah is no more a description of the creation of the earth and early Jewish life than the theory of instincts is a description of the psyches of a handful of bourgeois Viennese of the early twentieth century. The Torah is not the story of the Jews; it is what makes Judaism a story.

We tell myths for their own sake, because they are stories that insist on being stories—and insist on being told. We come to life at their touch.

However seriously we might regard them as so much inert poiema, and attach metaphysical meanings to them, they spring back out of their own vitality. When we look into a story to find its meaning, it is always a meaning we have brought with us to look at.

Myths are like magic trees in the garden of culture. They do not grow on but out of the silent earth of nature.. The more we strip these trees of their fruit or prune them back to our favored design, the more imposing and fecund they become.

Myths, told for their own sake, are not stories that have meanings, but stories that give meanings.

96

Storytellers become metaphysicians, or ideologists, when they come to believe they know the entire story of a people. This is history theatricalized, with the beginning and end in plain sight. A psychoanalyst who looks for the Freudian myth in patients imposes a filter that lets through nothing the psychoanalyst was not prepared to find.

The psychotherapeutic relationship will become horizonal only when both patient and therapist realize that the Freudian myth does not determine the meaning of what happens between them, but offers the possibility that their relationship will have a story of its own.

The Freudian myth does not therefore *repeat* itself in their relationship, but *resonates* in it. Those Jews who claim the right to certain territories on the basis of a biblical promise, those Christians who believe the Russians are the great evil armies foreseen in biblical prophecies of the end of the world, repeat the bible but do not resonate with it.

We resonate with myth when it resounds in us. A myth resounds in me when its voice is heard *in* mine but not heard *as* mine. I do not resonate when I quote Jeremiah or when I speak as Jeremiah, but only when Jeremiah speaks in a way that touches an original voice in me. The speech of New Yorkers resonates not because they talk like New Yorkers, but because when they talk we hear New York in their voice.

The resonance of myth collapses the apparent distinction between the story told by one person to another and the story of their telling and listening. It is one thing for you to tell me the story of Muhammad; it is quite another for me to tell the story of your telling me about Muhammad. Ordinarily we confine the story to the words of the speaker. But in doing so we treat it as a story quoted, not a story told. In your relating, and not repeating, the story of Muhammad, I am touched, and I respond from my genius. Something has begun. But in touching, you are also touched. Something has begun between us. Our relationship has opened forward dramatically. Since this drama emerged from the telling of the story of Muhammad, our story resonates with Muhammad's, and Muhammad's with ours.

As myths are told, and continue to resound in the telling, they come to us already richly resonant. The stories they are sound deeply with the stories of their telling. Their strength as stories lies in their ability to invite us into their drama. It is a drama that contains an entire history of voices, sounding and resounding from a thousand sources in our culture. For this reason myths are significantly unresolved—but unresolved in the way of an infinite game, having rules, or narrative structure, that allow any number of participants at any time to enter the drama without fixing its plot or bringing it to closure in a final scene. In such stories much will be said about closure, or death, but their telling will always disclose the way death comes in the course of play and not at its end.

97

Myths of irrepressible resonance have lost all trace of an author. Even when sacred texts are written down by an identifiable prophet or evangelist, it is invariably thought that these words were first spoken to their recorders and not spoken by them. Moses received the law and did not compose it. Muhammad heard the Quran and did not dictate it. Christians do not read Mark but the gospel according to Mark. Hindus understand their most authoritative texts, the Vedas, to be heard (*sruti*), and the literature that derives from the Vedas to be composed (*smriti*).

The gospel can be heard nowhere except from those who themselves have heard it. Although I might hear New York in your voice, there is no possibility I could hear New York by itself. No myth, therefore, exists by itself; neither does it have a discoverable origin. Whom could we name as the first New Yorker? Even when it is God who is heard by the prophet, it is a god who speaks in the language and idiom of the prophet, and not in locutions restricted to divine utterance—as though that god's speaking were itself a form of listening.

Indeed, myth is the highest form of our listening to each other, of offering a silence that makes the speech of the other possible. This is why listening is far more valued by religion than speaking. *Fides ex auditu.* Faith comes by listening, Paul said.

98

The opposite of resonance is amplification. A choir is the unified expression of voices resonating with each other; a loudspeaker is the amplification of a single voice, excluding all others. A bell resonates, a cannon amplifies. We listen to the bell, we are silenced by the cannon.

When a single voice is sufficiently amplified, it becomes a speaking that makes it impossible for any other voices to be heard. We do not listen to a loudspeaker for what is being said, but only because it is all that is being said. Magisterial speech is amplified speech; it is speech that silences. Loudspeaking is a mode of command, and therefore a speech designed to bring itself to an end as completely and swiftly as possible. The amplified voice seeks obedient action on the part of its hearers and an immediate end to their speech. There is no possibility of conversation with a loudspeaker.

Ideology is the amplification of myth. It is the assumption that since the beginning and end of history are known there is nothing more to say. History is therefore to be obediently lived out according to the ideology. Just as the warmakers of Europe regularly melted down the bells to recast them into cannon, the metaphysicians have found the meaning of their myths and announced those meanings without their narrative resonance. The myths themselves are now regarded as falsehoods or curiosities, and are therefore to be disregarded, if not forbidden.

What ideologists are concerned to hide is the choral nature

of history, the sense that it is a symphony of very different, even opposed, voices, each nonetheless making the other possible.

99

If it is true that myth provokes explanation, then it is also true that explanation's ultimate design is to eliminate myth. It is not just that the availability of bells in churches and town halls of Europe makes it possible to forge new cannon; it is that the cannon are forged in order to silence the bells. This is the contradiction of finite play in its highest form: to play in such a way that all need for play is erased.

The loudspeaker, successfully muting all other voices and therefore all possibility of conversation, is not listened to at all, and for that reason loses its own voice and becomes mere noise. Whenever we succeed in being the only speaker, there is no speaker at all. Julius Caesar originally sought power in Rome because he loved to play the very dangerous style of politics common to the Republic; but he played the game so well that he destroyed all his opponents, making it impossible for him to find genuinely dangerous combat. He was unable to do the very thing for which he sought power. His word was now irresistible, and for that reason he could speak with no one, and his isolation was complete. "We might almost say this man was looking for an assassination" (Syme).

If we are to say that all explanation is meant to silence myth itself, then it will follow that whenever we find people deeply committed to explanation and ideology, whenever play takes on the seriousness of warfare, we will find persons troubled by myths they cannot forget they have forgotten. The myths that cannot be forgotten are those so resonant with the paradox of silence they become the source of our thinking, even our culture, and our civilization.

These are the myths we can easily discover and name, but whose meanings continually elude us, myths whose conversion to truth never quite fills the bells of their resonance with the sand of metaphysical interpretation. These are often exceedingly simple stories. Abraham is an example. Although only two children were born to Abraham in his long life, and one of those was illegitimate, he was promised that his descendents would be as numberless as the stars of the heavens. All three of the West's major religions consider themselves children of Abraham, though each has often understood to be itself the only and final family of the patriarch, an understanding always threatened by the resounding phrase: numbered as the stars of the heavens. This is the myth of a future that always has a future; there is no closure in it. It is a myth of horizon.

The myth of the Buddha's enlightenment has the same paradox in it, the same provocation to explanation but with as little possibility of settling the matter. It is the story of a mere mortal, completely without divine aid, undertaking successfully a spiritual quest for release from all forms of bondage, including the need to report this release to others. The perfect unspeakability of this event has given rise to an immense flow of literature in scores of languages that shows no signs of abating.

Perhaps the Christian myth has been the narrative most disturbing to the ideological mind. It is, like those of Abraham and the Buddha, a very simple tale: that of a god who listens by becoming one of us. It is a god "emptied" of divinity, who gave up all privilege of commanding speech and "dwelt among us," coming "not to be served, but to serve," "being all things to all persons." But the worlds to which he came received him not. They no doubt preferred a god of magisterial utterance, a commanding idol, a theatrical likeness of their own finite designs. They did not expect an infinite listener who joyously took their unlikeness on himself, giving them their own voice through the silence of wonder, a healing and holy metaphor that leaves everything still to be said.

Those Christians who deafened themselves to the resonance of their own myth have driven their killing machines through the garden of history, but they did not kill the myth. The emptied divinity whom they have made into an Instrument of Vengeance continues to return as the Man of Sorrows bringing with him his unfinished story, and restoring the voices of the silenced.

100

The myth of Jesus is exemplary, but not necessary. No myth is necessary. There is no story that must be told. Stories do not have a truth that someone needs to reveal, or someone needs to hear. It is part of the myth of Jesus that it makes itself unnecessary; it is a narrative of the word becoming flesh, of *language entering history;* a narrative of the word becoming flesh and dying, of *history entering language.* Who listens to his myth cannot rise above history to utter timeless truths about it.

It is not necessary for infinite players to be Christians; indeed it is not possible for them to be Christians—seriously. Neither is it possible for them to be Buddhists, or Muslims, or atheists, or New Yorkers—seriously. All such titles can only be playful abstractions, mere performances for the sake of laughter.

Infinite players are not serious actors in any story, but the joyful poets of a story that continues to originate what they cannot finish.

101

There is but one infinite game.

INDEX

Numbers following entries refer to *sections*.